Medizinische Länderkunde
Geomedical Monograph Series

5

KENYA

Springer-Verlag Berlin · Heidelberg · New York 1978

Medizinische Länderkunde

Beiträge zur geographischen Medizin

Geomedical Monograph Series

Regional Studies in Geographical Medicine

Schriftenreihe der / Series of Monographs of the
Heidelberger Akademie der Wissenschaften · Mathematisch-naturwissenschaftliche Klasse
Begründet von / Founded by

Ernst Rodenwaldt †

Herausgegeben von / Edited by

Helmut J. Jusatz

em. ord. Professor Dr. med., Direktor des Instituts
für Tropenhygiene und öffentliches Gesundheitswesen am
Südasien-Institut der Universität Heidelberg i. R.
Leiter der Geomedizinischen Forschungsstelle
der Heidelberger Akademie der Wissenschaften

Unter Mitarbeit von / In collaboration with
Dr. phil. Berthold Carlberg †, wissenschaftl. Kartograph, Murnau/Obb. · Dr. rer. nat. Heinz Felten, Säugetierabteilung des Forschungsinstituts Senckenberg, Frankfurt/Main · Prof. Dr. med. Ludolph Fischer †, Tübingen · Prof. Dr. phil. Hermann Flohn, Direktor des Meteorologischen Instituts der Universität Bonn · Prof. Dr. phil. Gerhard Piekarski, Direktor des Instituts für medizinische Parasitologie der Universität Bonn · Prof. Dr. rer. nat. Ulrich Schweinfurth, Direktor des Instituts für Geographie am Südasien-Institut der Universität Heidelberg · Prof. Dr. phil. Drs. h. c. Carl Troll †, Bonn

KENYA

A Geomedical Monograph

by

H. J. Diesfeld

Professor Dr. med.

Direktor des Instituts für Tropenhygiene und
öffentliches Gesundheitswesen am Südasien-Institut
Universität Heidelberg

and

H. K. Hecklau

Professor Dr. rer. nat.

Fachgruppe Geographie des Fachbereichs III
Universität Trier

With 60 Photos, 17 Figures, and 9 Map-Plates

English Translation
J. A. Hellen, M. A. (Oxon.), Dr. phil. (Bonn) and Mrs. I. F. Hellen
Newcastle upon Tyne

Prof. Dr. H. J. Diesfeld
Südasieninstitut der Universität, 6900 Heidelberg

Prof. Dr. H. K. Hecklau
Fachgruppe Geographie, Fachbereich III der Universität, 5500 Trier

Additional material to this book can be downloaded from http://extras.springer.com

ISBN-13: 978-3-642-66937-8   e-ISBN-13: 978-3-642-66935-4
DOI: 10.1007/978-3-642-66935-4

Library of Congress Cataloging in Publication Data. Diesfeld, Hans Jochen. Kenya : a geomedical monograph. (Geomedical mono-graph series ; 5) Bibliography: p. Includes index. 1. Medical geography—Kenya. I. Hecklau, H. K., 1930– joint author. II. Title. III. Series: Medizinische Länderkunde ; 5. RA949.K4D53   362.1′09676′2   78-5516

Production of the Maps 1 – 9: Geomedizinische Forschungsstelle der Heidelberger Akademie der Wissenschaften. Printed: Karto-graphisches Atelier von Henning Wocke, Karlsruhe

Preface

The geomedical monograph on Kenya continues the series of Regional Studies in Geographical Medicine founded by Ernst Rodenwaldt in 1965, which may be considered as supplementing the World Atlas of Epidemic Diseases of the Heidelberg Academy of Sciences. Since the appearance of the first volume in 1967 the series has been edited by Helmut J. Jusatz under the aegis of the Mathematical and Natural Sciences Class of the Heidelberg Academy of Sciences.

The authors wish to express their thanks to the editor for his suggestion that they should undertake this kind of teamwork, and to acknowledge with gratitude the manner in which Professor Jusatz ensured the most harmonious cooperation, demonstrated his constructive understanding and made possible numerous specialist discussions both in small circles and in the Geomedical Symposia convoked by him in 1972 and 1974 at Schloss Reisensburg. The authors made use of this opportunity to present some methodological thoughts on the geomedical monograph under consideration here and to discuss them with specialists in the course of the second Reisensburg Symposium (Diesfeld and Hecklau, 1976, p. 72–88).

The authors regard the task of writing a geomedical monograph on Kenya as a continuation of their research begun in the mid-sixties as part of the Afrika Kartenwerk project of the Deutsche Forschungsgemeinschaft. They are obliged to Professor H. J. Jusatz, Heidelberg, and to Professor J. H. Schultze, Berlin, for entrusting them with the task of collating the material for the geo medical and the agrogeographical maps in the East Africa series. The Deutsche Forschungsgemeinschaft deserves the authors' thanks for its financial support of several prolonged periods of research in the countries of East Africa, without which it would have proved impossible to collect the requisite scientific material and to acquire the necessary knowledge of the area. In this context the authors wish to express their special thanks to the Deutsche Forschungsgemeinschaft for financing research in the field undertaken in 1974 in that part of Kenya which lay outside the Lake Victoria sheet of the Afrika Kartenwerk project.

The authors thank the Government of Kenya for granting permission to carry out the research and wish to express their appreciation to the authorities, civil servants, and colleagues in Kenya who helped to gather together the information required. Special thanks are due to Dr. Likimani, Director of Medical Services, Dr. Zal Onyango, Deputy Director of Medical Services, as well as to Dr. Koinange Karuga, Head of the Division of Communicable Disease Control, Dr. Itotia, Director of the National Public Health Laboratories, Dr. J. M. D. Roberts, Head of the Division of Vector-Borne Diseases, and to Mr. Ted Abukuse, Chief Statistician at the Ministry of Health, Nairobi.

The authors offer their particular thanks for support and valuable suggestions to Dr. L. C. Vogel, formerly Professor of Community Health in the Medical Faculty at the University of Nairobi and member of the Royal Tropical Institute, Amsterdam. Thanks are due to the Survey of Kenya for their willingness to allow the authors to reproduce some maps in the National Atlas of Kenya. The authors extend their thanks to their colleague, Professor S. H. Ominde of the Department of Geography, University of Nairobi, for his kind permission to use as a source the Population Distribution Map of 1969, which was produced in his department, and are equally obliged to Professor Ralph Jätzold, University of Trier, for making available the draft copy of his precipitation map of Kenya.

The Geomedical Research Unit produced a special base map of Kenya with a scale of 1 : 2 million for the cartographic presentation of the geomedical relationships between disease occurrences and geographical conditions. In this the orthography of place names follows that of the Kenya National Atlas, 3rd edition, Nairobi 1970. Conventional signs for settlement size classes were employed in accordance with the Kenya Population Census of 1969, Volume 3, Table 1.

The presentation of the results of the study was materially helped by the conscientious support of the staff at the Geomedical Research Unit of the Heidelberg Academy of Sciences and the Geography Group of Trier University. Herr H. Sauer of Heidelberg, as well as Messrs E. Lutz, E. Wagner, and D. Werle of Trier, collaborated in the production of the maps. The photographs were processed by Frau Wirtz of Trier and the typing of the manuscript was undertaken by Frau M. Albrecht of Heidelberg and Frau Kranz of Trier. Sincere thanks are extended to all of them. Special thanks also go to Dr. J. A. Hellen and Mrs. I. F. Hellen, Newcastle upon Tyne, for the translation of the text into English.

The authors thank Henning Wocke, the offset printers and cartographic studio, Karlsruhe, for their careful printing of the maps and the Springer Verlag for the good lay-out of the volume's text, diagrams, and photographs.

H. J. Diesfeld
H. Hecklau

Contents

Annex: Tables

Figures

         diseases:
         (see back of Map 9)
           1 Clinical Malaria
           2 Meningococcal Meningitis
           3 Hepatitis

                                                              4 Brucellosis
                                                              5 Pulmonary Tuberculosis
                                                              6 Acute Poliomyelitis
                                                              7 Tetanus
                                                              8 Anthrax

Maps

Methodological Considerations

The Geomedical Monograph of Kenya continues the series of geomedical monographs, but the authors of this volume attempt different approaches to the presentation of geomedical material. Apart from the more-or-less integrated geomedical theme proper, the preceding volumes on Libya (Kanter, 1967), Afghanistan (Fischer, 1968), Kuwait (French and Hill, 1971) and Ethiopia (Schaller and Kuls, 1972) all presented a unique and comprehensive documentation of the state of medical knowledge of the countries concerned, together with a largely complete bibliography of the medical literature. By contrast with the countries mentioned above, the medical, geographical, economic, and social-science literature on Kenya is already so prolific and internationally accessible that it does not seem appropriate to present a medical monograph which inclines to the encyclopaedic. A register of those diseases occurring in Kenya has been published recently (Vogel et al., 1974).

In view of the considerable literature on East Africa, an attempt has been made to structure the Kenyan geomedical monograph in the form of a medico-geographical analysis of the region and of a medico-ecological synopsis in the proper meaning of geomedicine (Jusatz, 1974; Diesfeld, 1974).

The authors endeavour to point out those correlations existing between the ecological endowment, the economic and social structure of the population, and the disease occurrences in the economic and socio-geographical regions of Kenya. The component regions and the region as a whole are related and compared with each other, and an evaluation of their development potential is made.

Earlier investigations, which were carried out in the context of the *Afrika Kartenwerk's* Lake Victoria sheet, have already shown that the physically and anthropo-geographically extraordinarily varied East African region is accompanied by an equally great diversity in the disease panorama (Diesfeld, 1969a; Hecklau, 1970). It was possible to differentiate so-called nosochores (i.e., sub-regions with typical disease patterns: Diesfeld, 1974) and to relate them as examples to the so-called biochores (agrogeographical types, after Hecklau, 1970) and geochores (as for example the "climato-types", after Jätzold, 1969).

So far as the epidemiological problem of method was concerned it became clear that the disease pattern observed at one hospital (i.e., a single point) could not simply be transferred to an entire area. Approached methodologically, the problem was contained with the aid of a terminological definition embracing the "population within a hospital catchment area" (Diesfeld, 1969a). However, it turned out that in many of Kenya's regions the catchment area could by no means be equated with an ecological region such as might have permitted conclusions about the relationship of ecology and disease.

Conceptually the Geomedical Monograph of Kenya proceeds from the definition of economic and socio-geographical regions. Moreover, the beginnings of an economic and socio-geographical division have already been set out by the mapping of agricultural land-use styles, since the methodological considerations involved in *Afrika Kartenwerk's* Agrogeographical Map of East Africa included the ecological and socio-economic functions of agricultural land use (Hecklau, 1976, 1978). The draft version of the economic and socio-geographical classification of Kenya will perforce ignore the more detailed differences among the styles of agricultural land use, such as small variations in the interactions of cultivated plants, whereas greater emphasis is placed upon such factors of the physical geography as morphology and altitude, soil and climate—in short, the ecological situation—as well as economic and socio-geographical circumstances, such as population density, social structure, and aspects of settlement and transport geography. The intention is to delimit the Kenyan state territory into regions, in which the living conditions of the population in their dependence on ecological and socio-economic factors within a certain spectrum present certain characteristics and differ from the living conditions of the population in adjoining regions. This should not of course lead to the assumption that in these regions of the country the conditions of life for the total population could ever be the same. Living conditions for the African population groups may have been relatively homogeneous in precolonial times, when farmers and herdsmen alike lived at the level of a pure subsistence economy. However, in the course of the country's economic development there has been a fairly pronounced differentiation of the formerly egalitarian farming and herding population, and in those areas in which large-scale pastoral farming, mixed farming or plantations established themselves, sharp social distinctions between Africans and non-Africans grew up. An attempt will be made to demonstrate that it is possible to distinguish regions in Kenya where the population presents a specific social structure, where distinct social groups can be recognised among the population. Moreover that these groups present typical forms of behaviour, which in turn affect the region and which, in their dependence upon the physico-geographical milieu, as well as the economic and socio-geographical conditions, also determine the style and standard of living of the groups in question. The division according to economic and socio-geographical criteria on the one hand, and according to physico-geographical ones on the other, presents an extraordinarily difficult problem, further aggravated by the fact that the physico-geographical conditions of the country are just as polymorphic as the economic and socio-geographical ones. The criteria used in such a division are investigated and chosen with the purpose of the regional differentiation in view. Providing a synthesis in the country's regions, the arrangement is intended as an aid to visual or conceptual representation of the multitude of physical and cultural geographical factors, insofar as they influence the living conditions of the population. The infrastructure of the health service and the density of distribution of its facilities in the regions are analysed as a special element. Supported by existing qualitative and quantitative information, the pattern of diseases and its relationship to ecological conditions in the biological field of nature is presented for each economic and socio-geographical sub-regions in the sense of Max Sorre's "Ecologie de l'Homme" (1947), of a geomedical analysis as defined by H. J. Jusatz (1974, 1976), and, in the field of sociology, in the sense of H. H. Barrows' "Human Ecology" (1923).

Instead of mapping medico-epidemiological data against the background of various thematic maps, the intention is to produce a geomedical-economic-socio-geographical synopsis, which, due to the advanced state of knowledge in the case of Kenya, ought to be possible for at least a number of variables.

The Geomedical Monograph will thus be divided into two major parts: an analytical foundation and a concluding interdisciplinary geographical-medical synopsis. The first part, not unlike a regional monograph, presents the ecological, demographic, settlement geographical, economic, social and transport geographical basis for regional division on economic and socio-geographical grounds. At the same time certain criteria for epidemiological, disease-causing or disease-restraining (i.e., medico-ecological) variables are introduced in order to direct the reader's attention to connections and reciprocal effects in the fields of medicine, geography, economics or social science.

In this section special attention will be devoted to the structural and functional analysis of the health services and to the presentation of the major diseases in their temporal and spatial dimensions, and these will be re-examined at a later stage in the geomedical context.

The synoptic section describes the patterns of diseases, causal connections, and reciprocal actions between disease and environment in the broadest sense of the word, which are characteristic of the sub-regions previously divided by economic and social spatial criteria.

In addition this chapter will recapitulate the regional density of medical facilities and the problem encountered in the implementation of the health services. Text, maps, and figures will carry equal weight in order to ensure a general understanding of the problem (see Diesfeld and Hecklau, 1976, p. 72 et seq.).

A. The Land and its Inhabitants

I. Fundamentals of Physical Geography

Kenya is a country of such great diversity in the regional sense, that the natural conditions of life of its population are extraordinarily varied. Kenya's palm-fringed coast along the Indian Ocean, with its fine-grained sand beaches and rich marine life, belongs to the earth's most renowned holiday areas. In the great national parks of the country the recollection of a pristine Africa, of the abundance of game in magically beautiful scenery, remains alive. The vast and gently undulating plains with their eye-catching inselbergen, and the varied forms of the dry savanna comprise the realm of pastoral nomads. Persevering and undemanding, they remain to roam the areas of scant precipitation, always on the move to new pastures and water holes, their existence constantly threatened by the loss of their herds when the rains fail for years on end, when grass and water-holes dry up. What a contrast this presents to the fertile, well-watered and densely settled Kenya Highlands inhabited by the small farmers. Numerous cultivated plants from Africa, Asia, America, and Europe flourish here: cereals and tubers, vegetables and fruit, coffee and tea. Healthy cattle thrive on the lush meadows.

The rainforests at higher altitudes, with their many species, remain almost untouched, although economic demands have led to the conversion of more and more woodland into farmland and coniferous forests.

In their upper reaches the montane forests are fringed by bamboo and heather growth, which gives way gradually to grassland and moorland. Almost exactly on the equator, the 5,199 metre high glaciated peaks of Mount Kenya overtop the country. After Kilimanjaro Mount Kenya is the second highest mountain in Africa. Like the other volcanoes of East Africa, its origin is linked with the formation of the Rift Valley. From the edge of the steep scarp slopes of the rift valley system the eye roves across breath-taking scenery: over savannas, salt lakes, picturesque extinct volcanoes, and also across the well-tended landscape of the large farms, providing maize, wheat, milk, and meat for the growing urban population of Kenya.

1. Surface Configuration

The diversity of Kenya's landscape is shaped by the relief, the formation of which can merely be outlined here. Plains and plateaux of varying altitude occupy the larger part of Kenya's surface area. Since the early Tertiary period the earth's crust has undergone several phases of epeirogenic uplift in Kenya. Some scholars support the view based on the Davisian cyclic theory that each phase of uplift was accompanied by the planation of the raised crustal section, commencing at the erosion base and extending to the vicinity of sea level. During the next phase of uplift the as yet uneroded parts of the peneplain were raised to a higher level so that a steplike erosion surface was formed. Ojany (1970, p. 4), for example, lends support to the view that there are remnants of six such important cycles of erosion:

"Altogether, relics of six important cycles of erosion (by pediplanation) can be recognised. The oldest and highest surface occurs as the skyline of many of the highest residual landmasses generally at above 2,000 m. It is well seen on the summit of Kisii Highlands and on the Cherangani, including the Trans-Nzoia, Elgeyo-Suk Plateaux, where it has been involved in considerable tectonic disturbances. The second surface is found from 1,800 to 2,000 m, and includes the summit of Kilungu and Mbooni Hills in Ukambani as the low bench in the Lerogi Hills. The third surface occurs between 1,500 – 1,650 m, and is well represented by the Kajiado surface as well as the surface upon which Machakos Town is built. The fourth surface is one of the best developed surfaces in the country. It occurs between 1,200 – 1,450 m. It is well developed in Siaya and Busia districts as well as over vast parts of Machakos and Kitui. One of the best examples can be traced in Masii-Kangondi-Tawa and Makutano areas of Machakos district. It is clearly the relic of the African surface of other authors. The last two surfaces occur at 900 – 1,200 m and 300 – 750 m above sea level. They are extremely extensive in the foreland plateau between the central highlands and the Coast. Both these surfaces are characterised by the occurrence of both the ubiquitous and marginal type inselbergs—whose summits sometimes prove to be remnants of the higher surfaces."

According to the representatives of the climatic-morphological school, the peneplains were simultaneously able to form or develop further and at different levels; those at the lower levels remained well preserved.

The peneplains, which attained an altitude of more than 1,500 – 2,000 m during the continuing uplift, met with cool and humid climatic regions where no planation occurs and were maturely dissected, as may be observed in the cool and humid altitudes of Kenya.

In Kenya's South West the peneplains have been uplifted several thousand metres, and at the peak of this upwarping the subsidence of the East African Rift System took place. This latter Rift System is part of that large fault-block depression which extends over 5,500 km in Africa from the Orange River in South Africa as far as the Afar Depression in Ethiopia and beyond the Red Sea into the Middle East. The origins and processes involved in the fault-block depression have not been unambiguously explained, and they now appear in a new light due to the findings of plate tectonics.

Between Lake Nakuru and Lake Naivasha, where the rift valley bottom reaches an altitude 1,900 – 2,000 m above sea level, the fault-block depression attains its maximum relief energy. The eastern flank of the rift is formed by the Aberdare Range, the highest peaks of which almost reach heights of 4,000 m above sea level. At the highest points of the Mau Escarpment the western flank of the rift nearly reaches 3,000 m above sea level.

In Kenya the first formation was accompanied by particularly marked volcanic activity, which has continued up to historic times. The Teleki volcano, south of Lake Rudolph, was last active in 1895. Mt. Longonot and Mt. Suswa manifest youthful forms too and can easily be viewed from the edge of the East African Rift Valley. However, Mt. Kenya, the highest peak of which—Batian—reaches 5,199 m above sea level, is only a ruin of a volcano, the more resistant vent filling of which has been exposed from the surrounding volcanic masses by erosion.

Further information and references to the literature on form and genesis of the relief may be found in L. Berry (1969) and W. T. W. Morgan (1973), to give but two examples.

Kenya's relief is of overriding importance for the climate of the country, particularly as regards the amount and distribution of precipitation and temperature. These climatic factors in turn condition the ecological endowment and the agrarian capacity, which greatly influence the density and distribution of the peasant and herding populations.

A comparison of the topographical map of Kenya (Map 1) with those of precipitation (Map 2) and population distribution (Map 4) shows the close connection between relief, altitudinal situation, the distribution of precipitation, and the distribution of population throughout the country; the distribution of population in an agrarian country corresponds much more strongly with the agricultural capacity—that is, its natural fertility—than in a highly developed industrial country. In the tropics far more than in the temperate zones the agrarian capacity depends more on the amount of precipitation than on the soil.

2. Climate

a) Precipitation: About two-thirds of the surface area of Kenya receives a long-term average of less than 508 mm (20 inches) of precipitation. These regions, poor in precipitation, cannot be agriculturally utilised without artificial irrigation. The poor ecological conditions permit only an extensive pastoral economy. The agrarian carrying capacity is low, and the population density itself is very low.

Those areas with 508 – 762 mm (20 – 30 inches) precipitation annually may be regarded as marginal areas for rainfall agriculture. Since they experience frequent crop failures on account of the paucity of precipitation, the nutrition of the population in these regions is therefore at times inadequate.

Areas with more than 762 mm (30 inches) precipitation are considered to be agriculturally fully utilisable. These areas are predominantly densely settled, except where they are covered by montane forest or are situated above the limit of agricultural cultivation.

Agricultural production, however, and with it the nutrition of the population, not only depends on the quantity of precipitation but also on its distribution throughout the year. But above all it depends on its reliability in setting-in every year. A glance at Special Maps A – D (Map 3) shows the great danger of drought-conditioned crop failures, and thus of famine, over large parts of Kenya.

As with the change of winter and summer in the temperate zones, so in the tropics the succession of rainy season and dry season determines the annual agricultural cycle and the rural population's life rhythm. Since Kenya is situated on both sides of the equator, some districts of the country enjoy two distinct rainy seasons, since these —as is well known—follow on the solstices. But this schematic statement holds true only for certain areas of Kenya. Since in some regions of the country there is but one rainy season, and on Lake Victoria precipitation occurs throughout the year, a clearly distinguishable rainy season cannot be determined. The air-masses of the land warm up more than those over the lake, with the result that local air-flows from lake to land take place. These lake winds, which are saturated with humidity, transport rain to the land at some distance from the coast of Lake Victoria throughout the year. In the permanently humid areas and in those with two clearly marked rainy seasons some crops can be harvested twice and others even three times in the year.

b) Temperature: Although Kenya is situated on both sides of the equator within the realm of the intertropical convergence zone, the hot, humid, and uniform climate with abundant rainfall, which is typical of the inner tropical lowlands, is greatly modified by the relief. According to the Köppen climatic classification, the higher altitudes of Kenya ought not to be regarded as tropical. Here diurnal temperature fluctuations are, however, greater than the annual ones. Thus due to the temperature the Kenyan climate is characterized as a tropical one in terms of C. Troll's definition.

Maps 3 C and 3 D show that temperatures in the densely populated upland locations are considerably below those of the lower situated and sparsely settled arid areas of Kenya. In the regions at above 1,500 m in particular the climate is well suited to Europeans, an advantage which favoured the establishment of European farm settlement in the former and the so-called White Highlands.

The gradation of temperature according to altitude has a decisive influence on the distribution of cultivated plants and the important tropical diseases. Thus in the distribution of malaria there is an upper limit, conditioned by temperature.

For further information and sources in the literature on climate see Thompson, B. W. and Sansom, H. W. (1967), Griffiths, J. F. (1969), Jätzold, R. (1977), and Morgan, W. T. W. (1973); see also National Atlas of Kenya (1970).

3. Hydrology

Most of the rivers which originate in the precipitation-rich Kenya Highlands run-off radially: numerous rivers, insofar as they do not dry out in the semi-arid areas or end up in seasonal swamps, merge with the river systems of the Tana and the Galana/Sabaki, which after a course of several hundred kilometres through semi-arid areas flow into the Indian Ocean.

Some rivers flow into Lake Victoria, as for example the Nzoia and Yala, others flow into areic lakes in the East African Rift Valley—the Kerio and Turkwell into Lake Rudolph, the Perkerra into Lake Baringo, to name only the most important of these. These areic lakes are alkaline, and Lake Magadi is even covered by a thick layer of soda, which is industrially exploited.

For the population, the waters play an important role. Since the construction of wells is still in its infancy, all rivers are of the greatest importance for the water supply of both, man and beast. Generally only the relatively clean mountain streams in the higher altitudes of settled areas are hygienically unobjectionable.

The water level of the rivers varies greatly. Whereas in the dry season they dry up or contract to a narrow channel, in the rainy season they swell enormously, and carry large amounts of sediment. Already during colonial period the administration began constructing earth dams in seasonally flowing rivers to check runoff during the rainy season, and thus conserve water for the dry season. Since man and beast alike tend to wade in and pollute the water, infectious diseases of various kinds are transmitted in these places (see Photo 39).

As rivers increasingly gain in importance for irrigated cultivation, they likewise serve to spread certain diseases like malaria and bilharzia into areas where these diseases have not yet occurred.

The rivers' possibilities for producing hydroelectric energy, however, are very limited. At present there are only two hydroelectric power stations of modest capacity, one on the upper Tana (Kindaruma Dam) and the other near Wanje, in the Murang'a (Fort Hall) District, on the Aberdare Range river (cf. Chapter IV, 4).

4. Ecological Potential

The criteria on which the divisions in Map 3 a, *Ecological Zones of Kenya,* were based, have been described by Pratt in the following terms: "In exact terms, these are eco-climatic zones, defined in terms of climate but described by reference to their climate and land use. Climate and especially rainfall is the primary factor which determines land potential in Kenya and it is climate which must therefore be given first consideration in planning land use. Furthermore, there exists a close relationship between rainfall and altitude, which is reflected in the vegetation . . ."

Ecological Zone 1 consists of "moorland and grassland or barren land, at high altitude above the forest line; of limited use and potential, except as a catchment and for tourism." Ecological Zone 1 is to be found only on Mt. Kenya and in the Aberdare Range. These are areas of a strange natural beauty of their own and have been designated National Parks. They are visited by numerous tourists.

Ecological Zone 2 consists of "forests and derived grasslands and bushlands, with or without natural glades. The potential is for forestry (with local wildlife and tourist development) or intensive agriculture, including pyrethrum, coffee and tea at higher altitudes. The natural grassland, under intensive management for optimum production, supports one stock unit per $1 - 1\frac{1}{2}$ ha., dependent on grassland type." With the exception of those areas which carry dense forest with a great variety of species, this zone belongs to the densely-settled, "active" areas of Kenya listed in Map 5 as a region with "peasant societies in areas of rainfall agriculture over 1,500 m above sea-level."

It is a zone in which the interests and demands of forestry and landscape conservation collide with the land requirements of the farming population.

Ecological Zone 3 and some areas at lower altitudes in Ecological Zone 2 provide the living area for the "peasant societies in areas of rainfall agriculture between 1,000 and 1,500 m above sea-level", and for the "African and Afro-Arabic peasant societies in areas of rainfall agriculture in the coastal lowland." During the colonial period, moreover, both these ecological zones witnessed the development of Afro-European class societies on mixed large-scale farms in areas of rainfed agriculture. *Ecological Zone 3* is "land, not of forest potential, carrying a variable vegetation cover (moist woodland, bushland and "savanna"), the trees are characteristically broad-leaved, e.g. Combretum) and the larger shrubs mostly evergreen. The agricultural potential is high, soil and topography permitting, with emphasis on ley farming. Areas under range-use are still extensive and under close management their stock-carrying capacity is high, at less than 2 ha. per stock unit."

The larger part of *Ecological Zone 4* accommodates Afro-European class societies engaged in pastoral farming; in parts it is populated by African peasant societies in marginal areas of rainfall agriculture and in other parts by African herding societies in predominantly arid and semi-arid areas.

Ecological Zone 4 consists of "land of marginal agricultural potential, carrying as natural vegetation dry forms of woodland and "savanna" (often as Acacia-The-meda association), or derived semi-evergreen or deciduous bushland. This is potentially productive rangeland — usually less than 4 ha. per stock unit — limited mainly by the encroachment of woody species. The more open country, with a high density of wildlife, constitutes a valuable tourist asset."

Ecological Zones 5 and 6 are characterized by Pratt in the following terms: Zone 5: "Land only very locally suited to agriculture, the woody vegetation being dominated by *Commiphora, Acacia* and allied genera, mostly of shrubby habit. Perennial grasses such as *Cenchrus ciliaris* and *Chloris roxburghiana* can dominate, but succumb readily to harsh management; more than 4 ha. is required per stock unit. Wildlife is important, particularly where dry thorn-bushland predominates. Burning requires great caution but can be highly effective in bush control".

Ecological Zone 6 is "rangeland of low potential the vegetation being dwarf shrub grassland, or a very dry form of bushed grassland with *Acacia reficiens* subsp. *misera,* often confined to water courses and depressions with barren land between. Perennial grasses (e.g. *Chrysopogon aucheri*) are localized within a predominantly annual grassland; productivity is confined largely to unreliable seasonal flushes and grazing systems must be based on nomadism. The populations of both wild and domestic stock are restricted severely by the environment".

Both these zones belong to the problem regions of Kenya, which carry a sparse population of "African herding societies." The Tsavo National Park is located in Zone 5; extending over 20,819 km^2, it covers an area about half the size of Switzerland. A glance at the map shows that *Ecological Zones 5 and 6* extend over about four-fifths of the Kenyan land surface. (The description of the ecological zones of Kenya by Pratt — see National Atlas of Kenya, 3rd ed., Nairobi 1970, p. 28.).

II. Demographic Basis

The population of Kenya comprises a multi-racial society, in which the Africans, according to the 1969 Census, account for 10.7 million persons, that is, 98.1% of the population. The African population is divided into tribal groups, which belong to totally different linguistic communities, as described in Chapter II. 1 a.

Europeans having their permanent residence in Kenya are also to be numbered with the Kenyan population. They totalled 40,593 persons in 1969, but represented only 0.37% of the country's entire population. Less than a tenth of them have adopted Kenyan nationality.

In the economic life of the country not only Europeans but also those population groups that have immigrated from the Indian sub-continent play an extraordinarily large role. In Kenya these immigrants are called "Asians", although they belong to the Indo-Germanic language family. Numbering 139,037 persons in 1969, they constitute 1.27% of the total population of the country. In 1969, 43.9% of them were Kenyan citizens.

Since the Middle Ages Arabs have settled on the coast of Kenya and in parts mixed with the Africans. In 1969 they numbered 27,886 persons, only 0.25% of the country's total population. Most of the Arabs in Kenya, that is 86.8%, are citizens of the country.

1. Development of Population

a) Africans: It may be supposed that Kenya, as most of the African countries, was sparsely populated in pre-colonial times. "The establishment of the Pax Britannica in the territory, the abolition of the slave trade, aid pro-grammes in cases of famine caused by pests or drought, the relief of famine by means of the introduction of pest and drought-resistant reserve crops, measures for the pro-tection of the health of man and animal— these and others have led to substantial increase in the population of East Africa over recent decades" (Hecklau, 1970, p. 64). In the 21 years between 1948 and 1969 the Afri-can population of Kenya has doubled, as is shown in Table 1.

Table 1. Population development in Kenya

Year	Africans	Asians	Euro-peans	Arabs	Others	Total
1911	?	11,787	3,175	9,100	99	
1921	?	25,253	9,651	10,102	627	
1931	3,000,000 (?)	43,623	16,812	12,166	1,346	
1948	5,252,753	97,687	29,660	24,174	3,325	5,407,599
1962	8,365,942	176,613	55,759	34,048	3,901	8,636,263
1969	10,733,202	139,037	40,593	27,886	1,987	10,942,705

Sources: Statistical Abstract 1975, Table 11.
         Kenya population census, 1962, Vol. I and II, p. 5; Vol. III, p. 2
         Kenya population census, 1969, Vol. I, p. 69.

At present the time taken for the world's population to double is estimated to be 35 years.

"The data on fertility and mortality ... indicated that the crude birth rate was probably between 48 and 51 per thousand, and the crude death rate between 18 and 25 per thousand. Thus the difference between the birth and death rates indicated rates of natural increase of be-tween 23 and 33 per thousand. However, the most plau-sible estimates for the early 1960s suggested a somewhat narrower range of between 27 and 31 per thousand or 2.7 and 3.1 per cent per annum" (Kenya population census, 1962, Vol. III. Nairobi 1966, p. 77).

Recent investigations into birth and death rates of the population of Kenya are not available. A comparison of census results of 1962 and 1969 leads to the assump-tion that the actual population growth amounted to slightly more than 3% during the period. According to a report submitted to the Government of Kenya by an ad-visory mission of the Population Council of the United States of America the population of Kenya will number 14.7 million inhabitants in 1980, 20.8 million in 1990, and 30.3 million at the turn of this century. (Family planning in Kenya. A report submitted to the Govern-

ment of Kenya by an advisory mission of the Population Council of the United States of America. Published by the Ministry of Economic Planning and Development, Nairobi 1967, p. 4).

b) Europeans: Kenya is one of the few African coun-tries in which a large number of white settlers established themselves. The immigration of British settlers to Kenya had several reasons. The first Europeans to visit Kenya were impressed by the lack of inhabitants in some regions of the Kenyan Highlands. It was soon recognized that the climate was suited to the permanent settlement of Europeans. However, the political decision to settle Ken-ya on the basis of a European farming economy was made because it was the fastest means possible of amor-tizing the Uganda Railway leading from Mombasa to Ki-sumu on Lake Victoria. This is discussed in detail in Chapter IV. 1.

The white immigrants, well versed in the rules of the game of a pluralistic society with a parliamentary demo-cracy, arranged for an area of about 3 million hectares to be reserved exclusively for acquisition by whites. This privilege they lost only in 1960, when about 400,000 hectares of farmland were repurchased from the Euro-peans and handed over to African settlers (cf. Chapter D. 1 e).

In population-geographical terms, development is significantly influenced by these economic and political processes. Until 1962 the number of Europeans had risen steadily to 55,759 persons. Since achievement of indepen-dence in 1963 the number of Europeans has steadily de-clined, reaching 40,593 persons at the time of the 1969 Census. But many more than 15,000 Europeans have left the country over these years, since others came to take their places. Only the roles of the whites changed, insofar as the Africans took over political power from the Euro-peans and replaced them in the leading positions in the administration. The newly arriving Europeans are either engaged in the economy or in advisory functions within the administration. They also play an important role in the health service: in 1964, 67% of medical practitioners were Europeans. As late as 1972, 63% of the doctors were still registered as non-citizens, and among the registered nurses 27.5% were non-citizens. In contrast to this, the term citizen includes a considerable number of natural-ized non-African people, Europeans and, above all, Asians (see Chapter B 6).

c) Asians: As already mentioned, in Kenya the immi-grants from the Indian sub-continent are known as Asians, some of whom have lived on the Kenyan coast for centuries, since the trading relationships between the Arab-ruled coast of Kenya on the Indian Ocean with Arabia and India reach back far into history. However, the immigration of Indian groups began on a large scale when Britain built the Mombasa-Kisumu Railway. It has been said, however, that after the completion of the pro-ject the majority of these workers returned home to their mother country, so that the real immigration of the Indian population groups set in only with the establishment of British colonial rule in East Africa. These immigrants came predominantly from Gujarat, the Punjab, and Goa.

Continuing immigration and a high birth rate caused the Asian population group to increase from 24,161 persons at the first census in 1911 to 176,613 persons at the census in 1969. The proportion of Asians born in Kenya thus increased from 32.1% in 1931 to 61.7% in 1962. Seventy-seven per cent of non-Kenyan born Asians came from India, 10% from Pakistan or Goa, and 12% from Kenya's East African neighbouring states.

According to calculations included in the Kenya population census it appears that in 1962 the birth rate among the Asian population scarcely exceeded 30‰ and that the death rate amounted to about 7‰. Compared with the African population this death rate among the Asians must be regarded as being very low. Several reasons account for this: the majority of the people concerned live under sound economic conditions; they are predominantly town dwellers, enjoying good medical care; and it may be that older Asians return to their home country.

When Kenya gained political independence, the Asian population group reached the 180,000 mark, slightly more than 1% of the total population in the country. But the Kenyan Asians occupied positions of economic supremacy, which effectively hindered the economic development of the African population. Kenya therefore is striving for the Kenyanization of the economy and administration. During the colonial period the Asians enjoyed the status of British protected persons. After the declaration of independence they were given the choice of either obtaining Kenyan citizenship or retaining their British passports. In December, 1967, a new law for foreigners, intended to promote the advancement of the indigenous workforce in Kenya, led to the first mass emigration of those Kenyan Asians who had opted to retain their British passports. Over the period 1968–1970 more than 25,000 Asians alone had to leave Kenya. According to the census of 1962 and 1969, the number of Asians in Kenya decreased from 176,613 persons to 139,037; therefore, more than 37,000 persons (=22%) must have emigrated, since the natural increase was not considered in this comparison (Kenya population census, 1962. Vol. IV, Non-African population, pp. 5–30; Kenya population census, 1969, Vol. I, p. 69).

d) Arabs: The Arab population group is made up chiefly of persons whose ancestors were already living on the East African coast in the Middle Ages, and many of whom have mixed with the African population. But in addition to these immigrants from the Arabian peninsula and the Near East also settled in East Africa during the colonial period. The distinction between "Kenyan Arabs" and "Arabian Arabs", as well as between Africans and indigenous Arabs, is often a difficult one. In the Lamu District on the coast, where the Bajun live, the distinction is particularly problematic after centuries of inter-

marriage. For this reason the Bajun are sometimes referred to as Africans and sometimes as non-Africans. Possibly it is a question of political opportunism when someone claims to be African or Arab at the time of a census. The statistical decline in the Arab population from 40,000 in 1962 to 28,000 inhabitants in 1969 may well be attributed to such problems of distinction. There was no emigration or natural regression in the population to account for the extent of the population decrease. In 1962, 89% of the 34,048 Arabs lived in the coastal region, above all on the coast itself. Fiftytwo per cent of this Arab population was concentrated in Mombasa itself, and it is unlikely that this pattern of distribution has changed much since that time. In 1962 the excess of births in the Arab population group amounted to 25 per thousand, with a birth rate of 40 per thousand and a death rate of 15 per thousand (Kenya population census, 1962. Vol. IV. Non-African population, pp. 57–68).

2. Distribution of Population

It has been estimated that almost nine-tenths of the Kenyan population live directly or indirectly from agriculture, and agriculture can flourish only where precipitation conditions permit. It is therefore not surprising that the narrow coastal strip on the Indian Ocean, which receives abundant precipitation, as well as those areas near Lake Victoria and at altitudes above 1,500 m are densely settled.

But the distribution of population not only depends on physical-geographical factors. Historical, economic, political, and social conditions modify the pattern considerably.

The eastern slopes of both the Aberdare Range and Mount Kenya together with the Kisii Highland and the Kakamega Hills are the most densely settled areas in Kenya. A traditional peasant population has been established in these areas for centuries. The density of settlement reaches values of 200–400 inhabitants per square kilometre; in some areas of the Aberdare Range and Mt. Kenya even this value is exceeded. The equally fertile and well-watered mountains inhabited by the Nandi and Kipsigis are, however, much more thinly settled, and here the population densities lie between 100–200 inhabitants per square kilometre. Both these tribes were formerly herding peoples, who changed over to more intensive cultivation only during the colonial period. The concentration of population among them has not progressed as far as among the Kikuyu, Meru, and Embu on the Aberdare Range and Mt. Kenya, or among the Kisii in the highlands of the same name, and Luhya in the Kakamega area. In pre-colonial times the peasant population was forced to settle close together in order to protect itself against the Masai raids.

The agriculturally utilisable areas of the former White Highlands are noticeably sparsely settled. Here there are values of less than 100, and in most districts even below 40 inhabitants per square kilometre. The cause is to be found in the European colonial govern-

Table 2. Population, area and density by provinces and districts, 1969

	Number	Land Area	Density *		Number	Land Area	Density *
	'000	sq. km.			'000	sq. km.	
Nairobi	509	684	745	**Rift Valley Province**			
Coast Province				Narok	125	18,513	7
Kilifi	308	12,414	25	Kajiado	86	20,963	4
Kwale	206	8,257	25	Nakuru	291	7,024	41
Lamu	22	6,506	3	Nandi	209	2,745	76
Mombasa	247	210	1,177	Kericho	479	4,890	98
Taita	111	16,959	7	Elgeyo Marakwet	159	2,722	59
Tana River	51	38,695	1	Baringo	162	10,627	15
Total	944	83,041	11	Turkana	165	60,824	3
				Samburu	70	20,808	3
North-Eastern Province				Trans Nzoia	124	2,468	50
Garissa	64	43,931	1	Uasin Gishu	191	3,784	50
Wajir	86	56,501	2	West Pokot	82	5,076	16
Mandera	95	26,470	4	Laikipia	66	9,718	7
Total	246	126,902	2	Total	2,210	170,162	13
Eastern Province				**Nyanza Province**			
Machakos	707	14,178	50	South Nyanza	663	5,714	116
Kitui	343	29,389	12	Kisii	675	2,196	307
Embu	179	2,714	66	Kisumu	401	2,081	193
Meru	597	9,922	60	Siaya	383	2,534	151
Isiolo	30	25,605	1	Total	2,122	12,525	169
Marsabit	52	72,732	1				
Total	1,907	154,540	12	**Western Province**			
				Kakamega	783	3,520	222
Central Province				Bungoma	345	3,074	112
Nyeri	361	3,284	110	Busia	200	1,629	123
Muranga	445	2,476	180	Total	1,328	8,223	162
Kirinyaga	217	1,437	181				
Kiambu	476	2,448	194				
Nyandarua	177	3,528	50	**Total Kenya**	10,943	569,249	19
Total	1,676	13,173	127				

* Persons per square kilometre.
Source: Statistical Abstract 1975, Table 13

ment's settlement policy, which, as already mentioned, reserved the acquisition of land in the White Highlands for Europeans. Nevertheless, there was illegal settlement by the Africans in these areas, which will be taken up later. At present these areas are quickly being filled, especially where formerly European-owned farms are being divided and placed in African hands (cf. Chapter D. 1 e). The coastal area is densely settled only in parts; only in the urban catchment area of Mombasa does the populations density exceed of 500 inhabitants per square kilometre. The remaining areas show population densities at times considerably below 100 inhabitants per square kilometre. These constitute land reserves which are being put into agricultural use (cf. Chapter D. 1 c and D. 1 e and Table 2).

Areas marginal for rainfall agriculture, receiving about 510 – 760 mm precipitation, frequently suffer from crop failure due to drought and consequently are very sparsely populated. Large areas of the Eastern Province with less than 40, and in parts even less than 20, inhabitants per square kilometre are numbered among these.

The most sparsely populated areas are those in which a pastoral subsistence economy is practiced. These are predominantly regions in which, for economic reasons, agriculture can be engaged in in only a few specially favoured places.

Here the population density does not exceed 10 inhabitants per square kilometre in any district, and in some areas it is even lower than one inhabitant per square kilometre.

3. Structure of the African Population

The overwhelming majority of the African population of Kenya consists of sedentary peasant farmers. Though members of different racial and linguistic groups, their life styles and living standards are similar. Only in more recent times has there been a tendency to a greater differentiation of peasant society through the commercialization of agriculture. The way of life of the herding population, which accounts for less than one-tenth of the population, is totally different however. Some fringe groups with aboriginal ways of life who gain their livelihood as primitive hunters, gleaners, and fishermen are numerically of little importance. Statistically so far unrecorded, and therefore not verified by numerical data, is the more recent social and occupational differentiation of the population into those classes which are above the peasant farming and herding society in the social hierarchy. They form those population groups employed in non-agrarian branches of the economy, administration, politics, science, and cultural affairs, etc. Undoubtedly the African elite has grown up which has its function in all these spheres of activity.

a) *Ethnic differentiation:* Kenya lies where the distribution areas of the Bantu, Hamites, Nilo-Hamites and Nilotes converge. Moreover, for centuries, as has already been noted, the coast has come under the Arab zone of influence. These are the causes of the marked heterogeneity of the African population, through its membership of various racial and linguistic groupings, shown in Table I (Annex) and Map 4a. A differentiation according to racial and linguistic characteristics is, however, difficult because of the centuries-old intermingling between the individual tribes. In the Kenya Population Census, 1962, Vol. III, p. 34, it is noted that: "For the purposes of the 1962 Census, the indigenous African and Somali population of Kenya was grouped into 40 tribes... The indigenous tribes were divided into eight major groups on the basis of ethnic, linguistic and geographical considerations... The grouping is of course somewhat arbitrary in many cases."

The heterogeneity of the Kenyan population constitutes a difficult problem. Again and again the overcoming of tribalism is proclaimed, but neither the contrasts can be ignored nor the competing demands for the wherewithal for development in particular tribal regions of influence at a supra-regional level.

Without exaggeration it may be claimed that in Kenya the Bantu — in particular the 2.2 million Kikuyu — represent the most influential population group. The Luo (1.5 million), living on the shores of Lake Victoria, are the most powerful group of non-Bantu peoples. Among both these population groups conflicts of interest can assume rather sharp forms. At present the Bantu tribes, farming by far the largest part of the agriculturally utilisable soils of Kenya, represent 65.5% of the country's total population. They are assumed to have moved into their present settlement areas between the 17th and 19th centuries, yet the areas of departure of these Bantu migrations to Kenya remain a moot point. Although the Nile-Congo watershed has been regarded as a possible area of origin of the Bantu groups, most of them trace their own origin from those tribes who lived on the shores of the large Ugandan lakes (National Atlas of Kenya, Third Edition 1970, p. 50).

The Nilotic Luo migrated from the western part of the southern Sudan to Uganda and western Kenya in pre-colonial times; their present tribal area extends from Lake Victoria to northern Tanzania. Numbering 1.5 million members, they represent 14% of the Kenyan population. Pressure for land and an unfavourable agrarian structure, which have so far considerably complicated improvement in agriculture, have caused many young and active Luo to seek work outside their home districts. The Nilo-Hamitic tribal groups, which migrated from the southern Sudan, inhabit the extensive area of the East African Rift Valley and the western frontier zone of northern Kenya. The Turkana are their main representatives. The Masai are, however, among the best-known herding tribes of Kenya. In pre-colonial times they ruled over a large territory in East Africa. Disease and famine caused by the loss of their cattle greatly diminished their importance towards the end of the last century. In 1902 and 1912 the colonial administration concluded two much disputed agreements with the Masai, under which the Kenyan Masai were obliged to restrict themselves to their present area, the Kajiado and Narok districts, thereby vacating land for European settlers in the former White Highlands. By far the most numerous part of the Nilo-Hamitic population group lives in areas of pastoral economy with agriculture feasible in only comparatively small areas. They pursue a pastoral subsistence economy with only sporadic marketable production. In those few areas where ecological conditions permit agriculture, entire herding tribes, like the Nandi and Kipsigis, have changed over to cultivation. Among the Elegeyo, Marakwet, and Pokot, those families living in areas suitable only for pastoralism remain completely as herders, while others in the more ecologically favourable areas have likewise adopted sedentary cultivation.

The Hamitic tribes who inhabit the arid north east quarter of Kenya are also considered the purely nomadic herdsmen of the country. For largely ecological reasons and because of the extremely sparse population density of the vast territory, they have hardly participated in any modern development. They are assumed to form part of the final Hamitic migration, which moved from Ethiopia via Somalia into Kenya. In the Sixties, efforts were made by the Somalis to annex their territory in Kenya to that of the Republic of Somalia. The international frontier has divided the tribal area of the Somali in the same way as the tribal area of the Masai in the south of Kenya.

b) *Age structure:* The age differentiation for Kenya's population is typical of that of all African countries and for the developing countries at large; almost half — that is 48.4% of the population — is under 14 years of age. In Sweden, which may serve as an example of a developed industrial country, the child population under 14 years of age amounts to only 20.7% of the population as a whole. In Kenya only 5.4% of the population are to be found in the over-sixties age group, compared with 21.4% in Sweden. The age group of persons capable of employment, which amounts to 53.8% in Kenya, is lower in Kenya than in Sweden, where it comprises 57.9% of the population (see Fig. 1 on back of Map 2).

There are differences in the extent of illiteracy among the sexes. In all age groups the number of those neither able to read nor write is lower among males. In the age group 15 to 19 years for example, about 60% of the females, as opposed to 40% of the young males, are illiterate. The short life expectancy for the population of

the developing countries which emerges from these figures has serious consequences for the development of the
country.

The proportion of the potential working population
between the ages of 14 and 65 is smaller by comparison
with the industrialised countries as mentioned above.
Since due to improved medical care among the population many more children will reach reproductive age in
the future than did in the past, these conditions will
change. A further population explosion may be envis-
aged.

Increasing improvement in life expectancy will result
in smaller losses in each cohort. Even the number of
births may one day decline. According to the *historico-
sociological population law in the industrial age* formulated by
Mackenroth (1953, p. 128), in regard to its population
growth Kenya has reached the point when a demographic change-over begins to take effect, namely one which
is characterised by a high birth rate and falling death
rate. It is a question of great importance for the country
whether the improvement of economic and social conditions will be successful enough to permit the adoption of
the procreative behaviour of the industrial nations, which
is characterised by a low birth and low death rate. In
other words, in Kenya as in all developing countries the
balancing of the economic growth rate and the development of social conditions in agreement with that of the
population growth rate is of critical importance.

4. Internal Migrations

Internal migrations of individuals and families, by
contrast with the migrations of entire tribes, were only
set in motion when the country's economy began to develop and to differentiate itself. Two migratory movements arose: the migrations from rural areas into the
towns and those from densely settled peasant farming
areas into others with large farm agriculture. The large
plantations in particular require great numbers of workers. E. H. Ominde has described the population movements in detail in his book "Land and Population Movements in Kenya" (1968).

The areas of origin are the densely settled peasant
areas in the Central Province and in the former Nyanza
provinces of western Kenya. The immigrants into the
former White Highlands applied for work on the European farms or settled illegally in these areas as so-called
squatters. Since the Europeans depended on African labour for the operating and development of their farms, in
1962, at the height of the development of the European
farms in the Trans-Nzoia and the Uasin Gishu district,
there were about 200,000 Africans compared with only
2,500 Europeans and 6,000 Asians living in this area. The
concentration of settlement took place chiefly in those
areas in the course of the "One Million Acre Settlement
Scheme 1962 – 1966", under which formerly European
farms were shared out to African settlers, above all in the
Nyandarua and Uasin Gishu Districts, as well as to a

limited degree in the districts of Kericho, Trans Nzoia,
and Nakuru (see Map 5).

Migrations from the countryside to the town are the
expression of far-reaching changes in the economic and
social structure of Kenya. The most important movement into towns is the migration from all parts of the
country to Nairobi, followed by that to Mombasa. Movements to the remaining towns of the country are of a
much lower order.

Population growth and the increasing scarcity of farm
land are forcing more and more Africans to leave their
home districts in order to find work in the towns. Even
though all the Africans arriving in towns by no means
succeed in finding employment, the flight from the land,
together with the overall economic development of Kenya, has led to a considerable growth of towns, as shown
in Table 3.

*Table 3. African population growth in the main urban settlements,
1948 – 1969*

Town or City	Inhabitants			Percentage increase 1948 – 1969
	1948	1962	1969	
Nairobi	64,397	156,246	509,286	690,85
Mombasa	42,853	111,847	247,073	476,56
Nakuru	12,845	30,189	47,151	267,08
Eldoret	5,408	15,059	18,196	236,46
Kisumu	5,336	14,119	32,431	507,78
Thika	2,806	11,352	18,387	555,28
Nanyuki	3,041	8,919	11,624	282,24
Kitale	4,344	7,000	11,573	166,41
Nyeri	1,858	6,256	10,004	438,43
Kericho	2,243	5,950	10,144	352,25

Source: Kenya population census, 1962. Vol. III. African population, p. 23.
Kenya population census, 1969, Vol. II. Data on urban population, p. 1.

Internal migration from rural to urban areas creates
considerable social, economic, psychological, and medical
problems. It leads to a distinct shift in the sex ratio. In
the towns the disruption of the family leads to increased
promiscuity, with all its social and health risks. Even
assuming that the data are reliable only to a certain extent, the 1969 Census reveals this problem quite clearly.
The overall sex ratio for Kenya is $\male : \female = 1,000 : 1,028$
for adults (> 16 years of age) and $1,000 : 956$ for children (< 15 years of age). Without going into the particulars of the childhood sex ratios, the proportion of adults
to children may, in accordance with the absolute census
figures and the above definition, be taken as $1,000 : 1,020$,
with the relationship among males amounting to $1,000 :
1,053$, and among females to $1,000 : 1,012$. The adult to
child sex ratio index of 1.06 for the whole of Kenya
provides an indication of the average, natural, sex-specific
relationship of adults to children. As is to be expected
the urban centres of Kenya are seriously disrupted in this
respect, especially in Nairobi, Mombasa, and Nakuru.

Whereas the childhood sex ratio is to be regarded as balanced, it is massively displaced in adulthood. In the rural areas "undisturbed" provinces can be identified (see Table 4) by their having in the main a balanced sex ratio and sex ratio-age index between 0.96 and 1.06, i.e., having a small difference between adult and child sex ratios. This situation is prevalent in the Coast, North Eastern, and Rift Valley Provinces. By contrast, the Central, Eastern, Nyanza, and Western Provinces present markedly disturbed sex ratios, with a significant prevalence of the female component and a marked difference between adult and child sex ratios (a sex ratio/age index around 1.2). These demographic calculations throw light on a problem, that would appear even more striking in any detailed presentation at the district level.

Another aspect of the problem emerges from the tribal classification of the population in Nairobi and Mombasa according to the 1969 Census (see Fig. 2 and 3).

Such a statistical interpretation of the census will, however, have to bear in mind that these figures comprise all age groups, including children having as a rule an undisturbed sex ratio.

In Nairobi almost 47% of the Kenyan Africans are Kikuyu; they are followed by the Luo, Luhya, and Kamba with about 15% each. All the other tribes together number about 7%. Among the Kikuyu, the sex ratio of 1,000 : 692 is the one showing the least distortion; the Kamba tribe with a ratio of 1,000 : 472 has the most disturbed sex ratio among those living in Nairobi.

In Mombasa only 34.3% of the population stems from the surrounding district, and has in consequence experienced the least disruption, in a sex ratio of 1,000 : 715. The Kamba with 17% and a sex ratio of 1,000 : 524 are the group next to them in indigenous strength, but the population of the surrounding of Mombasa suffer most from the unbalanced sex ratio. Luo and Luhya are the groups following which have a relatively favourable sex ratio. These data clearly indicate that the problems connected with internal migration are certainly more complex than is generally assumed, not only in the urban areas but also in the areas of out-migration. Resettlement areas on the other hand receive whole families—a fact well substantiated by the census data on the Rift Valley districts.

Table 4. Male : female ratio in adults and children in urban centres and in rural provinces of Kenya

	Male : female – ratio :		
	Adults 1,000 males per	Children 1,000 males per	Adults-Children-Sex ratio – Index (ACSI) *
I. Urban Centres			
Nairobi	537 female	1,008 female	0.53
Mombasa	621	914	0.68
Nakuru	665	1,016	0.65
Total Kenya urban Centres	590	988	0.60
II. Rural Provines			
Central	1,171	984	1.20
Coast	953	956	1.00
Eastern	1,172	972	1.21
North Eastern	858	817	1.05
Nyanza	1,109	956	1.16
Rift Valley	923	957	0.96
Western	1,187	981	1.20
III. Kenya total	1,028	956	1.06

Source: Compiled from Kenya Population Census, 1969

$$* \text{ ACSI} = \dfrac{\dfrac{N° \female \text{ adults}}{1,000 \male \text{ adults}}}{\dfrac{N° \female \text{ children}}{1,000 \male \text{ children}}}$$

ACSI ∼ 1.0
ACSI ≪ 1.0 = prevalence of male
ACSI ≫ 1.0 = prevalence of female

Nairobi

Population by Sex and Tribe, 1969

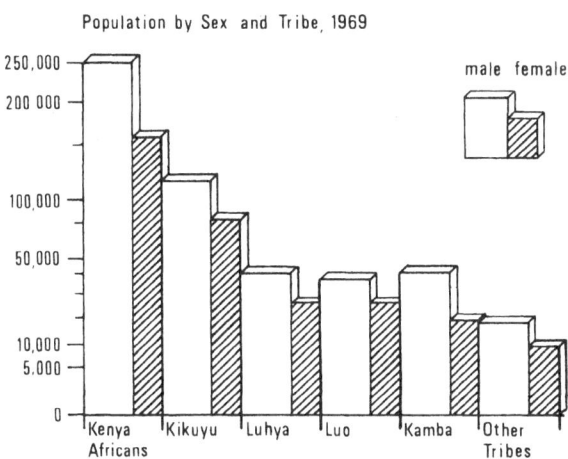

Source : Kenya Population Census 1969, Vol. II. Data on Urban Population. Table 3, p. 3 f.

Figure 2.

Mombasa

Population by Sex and Tribe 1969

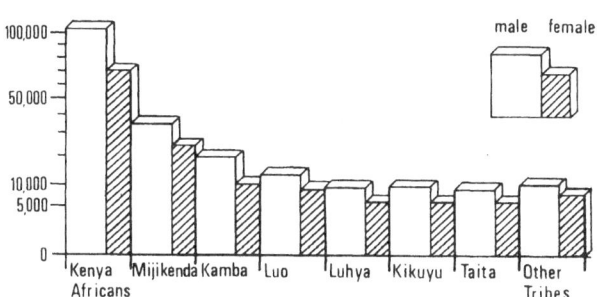

Source : Kenya Population Census 1969, Vol. II. Data on Urban Population. Table 3, p. 3 f.

Figure 3.

III. Fundamentals of Settlement Geography

1. Rural Settlements

East Africa lacks villages in the European sense of the word. The African peasant farming and herding population prefers the isolated or dispersed settlement in which families with their dwellings keep at a certain distance from their neighbours. A typical peasant settlement consist of several dwelling huts, the number of which depends on the number of wives and children of the head of the family. The various Kenyan tribes have developed different hut forms, the originally dominant form being that of the one-roomed round hut, the walls of which are constructed with poles and branches and covered with mud. Poles jointed to form a conical apex provide the framework for the roof, which is interwoven with rods, tipped on to the walls of the hut and covered with grass. Among only a few tribes, for example the Pokomo, hut types predominate which have no wall construction, but rather roof and side-walls are plaited in one piece and covered with grass. In recent times more and more peasant families have built multi-roomed, rectangular houses from traditional materials described above (Photo 18). The somewhat more affluent peasant families substitute corrugated iron sheeting for grass on their roofs. Successful farmers—the so-called progressive farmers—and families with correspondingly high cash incomes have solid, multi-roomed houses built for them, complete with corrugated iron roofs and glazed windows (Photo 22).

As a rule the dwellings made of mud walls and grass roofs are not free of vermin, such as rats, mice, and harmful insects of various sorts, which serve as vectors of diseases. The transition to corrugated iron roofs and solid construction has brought about a reduction in the transmission of infectious diseases. Mainly this, as well as general change in house forms, and improvement in turn of hygiene conditions in the settlements have reduced and sometimes almost eliminated certain diseases such as plague, tick typhus, and tick-borne relapsing fever, which once spread epidemically. The reason for this may be seen in the general endeavour in recent years to improve the living standard as well as health education. The dwelling place of a peasant family is complemented by several storehouses for the safekeeping of crops. In principle these storage huts frequently consist of plaited baskets about 1½ – 2 m in diameter, which are covered with mud. The stores are capped by conical grass roofs, which are lifted off when the harvested crop is deposited or removed. For protection against pests the stores are placed on stones or a low pedestal of branches (Photo 20).

Due to the bad air-flow through the storage bins part of the crop frequently becomes mouldy if it has not been kept in a state of perfect dryness. Damage is also caused by rodents which manage to intrude in spite of the preventive measures outlined above.

The rural settlement places of so-called progressive farmers, who, thanks to cash income and credit, can afford the cost of erecting a modern farm, have undergone a complete transformation: besides the modern dwelling described above, primitive storage bins have been replaced on well-built progressive farms by stalls for cattle, buildings for fodder and foodstuffs, a tick-spraying facility, corrugated tanks for collecting rainwater, and quite frequently a little flower garden (Photo 33).

The peasant farming population had been encouraged during the colonial period to make so-called pit latrines in order to restrict the transmission of intestinal parasites. However, their effectiveness depends upon their being regularly cleaned, and this is guaranteed best if a separate latrine is available for each family. Thus, the common practice of communally used latrines, found particularly in the labour compounds of the plantations, serves more to propagate than to reduce parasites, since these conditions are not conducive to cleanliness.

While the peasant population of Kenya is observed to have brought far-reaching changes in the overall picture of rural settlement, most of the herding tribes in Kenya have scarcely altered their original settlement forms. As a rule, one extended family uses a common settlement, although sometimes several extended families live together. Even at the present time, among many of the pure pastoral nomads a tent is preferred; this is packed on the camels and transported from one pasture to the next. Many herding families do have, however, an established base where the old people, women, and children remain while the persevering men are away with some of their animals; in other cases such herding families frequently move from base to base, with the effect that each of these is only temporarily inhabited. In some ecologically well-endowed areas some of the herdsmen have already become sedentary, and some have even adopted modern forms of pastoral economy, including the buildings that go with it.

The boma or manyatta consists of thorn bushes piled up in a circular fashion surrounding the dwelling huts and the kraal for the livestock. The settlements of the Masai, probably the best known East African herding tribe, are particularly characteristic. The Masai hut consists of a box-shaped woven structure of branches as thick as an arm, covered on the outside by a mixture of earth and dung. The hut is partitioned into several small rooms, which serve as bedrooms and kitchen. All the rooms are without windows, and the smoke from the fireplace passes through the entire hut, so that the woven material is blackened by soot. Only above the fireplace is there a hole in the ceiling, and this about the size of a fist; this allows some smoke to escape and some light to penetrate. The room next to the entrance houses calves and heifers at night.

Cattle are milked in front of the hut and are kept very close to it overnight. The ground in the manyattas is covered by cattle dung, and man and beast are troubled by innumerable flies. Living together in crowded conditions in poorly ventilated, smokey huts, under the worst possible conditions of hygiene and with chronic and extreme shortage of water, promotes the transmission of a number of infectious diseases such as tuberculosis and other respiratory diseases, chronic eye infections, meningitis and, among children particularly, scabies and purulent skin infections. Life in the huts' environs, fouled as they are by cattle excreta, is exposed to the transmission of tetanus, anthrax, brucellosis and salmonella.

2. Urban Settlements

Towns founded by the Arabs along the Kenyan coast date back to the Middle Ages and of these Mombasa developed into East Africa's most important port during the colonial period, while Malindi and Lamu have become centres for the long-distance tourist industry. The towns of the interior of Kenya were recently established by the colonial government after the turn of the century. Some of them have experienced an astonishing development, above all Nairobi itself, set up in 1899 as the headquarters of the railway administration. In 1907 it became the seat of the British administration for the protectorate, and today it has 600,000 inhabitants, 70,000 of whom are Asians and another 19,000 Europeans. As the country's capital, the place where numerous firms with world-wide connections are represented, Nairobi can boast a central area that is modern in both its buildings and functions. Public buildings like the Parliament, government buildings, city administration, foreign embassies, and high-rise blocks housing business representatives are concentrated here, and adjoining these is a first-class shopping area with shops catering for higher order goods and services. A second class shopping area, in which a large range of reasonably-priced goods is on display, completes the central area in the north east.

Nairobi is a cosmopolitan city drawing its inhabitants from three continents and witnessing tourists and business and commercial travellers arriving from all over the world in tens of thousands each year. With the construction of the ultra modern Congress Centre, Nairobi, under the title of "City in the Sun", is striving to become a city of international congresses. The multiracial and social differentiation of Nairobi's inhabitants finds its expression in the formation of residential quarters. Nairobi's best and most expensive residential area —called Upper Nairobi by Halliman and Morgan (1967) — is situated at an altitude of about 1,700 m above sea level to the west and north of the city centre. In the early sixties about 80 per cent of the Europeans in Nairobi lived there, as well as Africans and Asians in the higher income groups.

"The great majority of Asians live in the Parklands-Eastleigh area, with another group in the recently established Nairobi South district" (Halliman and Morgan, 1968, p. 106).

More than two-thirds of the African population live in Eastlands, a district in which the City Council and big employers have erected working-class estates (Halliman and Morgan, 1968, p. 186).

The industrial area of Nairobi, conveniently situated near the railway and airport in the east of the city, has become one of the most important industrial locations in East Africa. In the early seventies Nairobi's railway workshops alone provided employment for more than 7,000 workers.

Nairobi is also the cultural centre of the country with agencies of publishing houses, two theatres, and a fast-developing university. In the early seventies about 4,000 students were enrolled at Nairobi University, 1,000 of whom were non-Kenyans. The Faculty of Medicine was opened in 1967, and Nairobi City Council has its own Health Department. For medical institutions belonging to the university see Chapter B.

3. Central Place System

The central place system is best developed in the Large Farm Areas of the former White Highlands. Nakuru, located in the Rift Valley, has achieved the position of the main regional centre for the Large Farm Areas. It is the focus of some urban functions the extent of which can be traced from the Rift Valley via the farming areas in the Mau Forest to the Uasin-Gishu and Trans-Nzoia Plateau. These include administrative establishments, manufacturing industries, and marketing organisations concerned with the agricultural products of the large farms. Nakuru also acts as a shopping centre for the more select needs of the farming population (Photo 56). Moreover there are some small country towns in the Large Farm Areas, which might possibly be described as medium-order centres, and these include Kitale, Eldoret, Naivasha, Nanyuki, and Thika. Thomson's Falls, Gilgil, Elburgon, Njoro and Molo ought to be described as low-order centres. The medium-order centres are all located along the railway line and connected by well-made asphalt roads. In general it may be said that the centres' topographical position is indicative of a well-considered choice of location, determined by the requirements of the agrarian *Umland*. The layout of the towns shows the regulating and planning hand of the town planner. The formation of distinct quarters was the intention of the planners and is an expression of the strict segregation of African, Asian, and European population groups during the colonial period. As regards the population and the manner of construction, each quarter is characterized by a large degree of conformity in itself, but compared with the other quarters they often show great variation. In some towns they are spatially far apart from one another. The formation of these quarters is an expression of racial and social grouping in the population as it developed before Kenyan independence, and to a certain degree it still exists. Even at the present time barriers between these different population groups persist. Although informal social contacts are increasing between these groups, family connections between members of the different groups remain rare. Each group fulfills certain economic and social functions, and it is only since independence that here, too, the strict confines have become blurred. Before independence the apex of the social hierarchy was reserved exclusively for Europeans; in the towns of the Large Farm Areas they occupied positions as administrators, as scientists on the experimental stations, in independent professions, as business people and managers in various capacities. The Europeans lived in residential quarters made up of large plots with generously designed country houses. Now the young African elite is pushing its way into positions formerly held by Europeans, and this elite now lives in what were the European-owned quarters. Between the mass of the African population, and the Europeans, and the new African elite, the middle class is formed by Indians predominantly engaged in trade, commerce, and in the public services. They do not present a uniform group but are split into different religious communities, which are, in addition, greatly differentiated according to property and income.

The lowest social class is composed of the Africans who arrive in the towns without means or education and provide a labour reservoir for all the lower services, although very gradually, even here qualifications for skilled occupations are bringing into existence groups with higher earning capacity. Strict supervision of immigration has succeeded in preventing the formation of slums in the central places of the Large Farm Areas. Those Africans of the lowest income groups live in clean quarters, laid out according to plan; some of these are small but solidly built uninterrupted terrace or serial houses, others—similarly constructed—are round houses modelled on the African hut.

By contrast with towns of similar size in Europe, the towns of the Large Farm Areas function almost exclusively as centres for the agrarian *Umland*. They act as collecting places for the agrarian produce, for storing, marketing and processing it, or transporting it to places for further processing or sale (see Hecklau, 1968, p. 182 et seq.).

The central place systems in areas of peasant farming and cattle keeping in Kenya have undergone development to varying degrees. As provincial capital of the Coastal Province, Mombasa is not only a regional centre for the population of the coastal region, but as a commercial- and industrial port-city it is also the seat of multinational firms having commercial relations which extend far beyond Kenya into the East African hinterland.

Kisumu on Lake Victoria has also reached a level of economic development more like that of a major centre of supra-regional importance. As the terminus of the Mombasa – Kisumu Railway, it is an important trading centre and entrepôt on Lake Victoria.

By contrast, the main centres in the principal African farming areas often fail to attain even the size of a small town, although they accommodate the provincial or district administration, and at times even both. The centres are in the main of local functional significance for the smallholding or herding population concerned. They are the seat of administration, of trade, and of commerce; such centres have at their disposal schools and religious and medical institutions of varying importance.

In the African smallholding and herding areas the towns were founded by the European administrators during the colonial period, although never more than a small number of Europeans resided in them. These latter were employed in the administration, on mission stations, or in the education and health services. Urban life in the form of trade and commerce was developed by the Asian population group. Asian traders penetrated the remotest corners of the country where they offered their goods or exchanged them for the agrarian products of the Africans. They have thus made an important contribution to the commercialisation of Kenyan agriculture and, at the same time, have laid the foundation for genuine urban development.

At first the Africans remained cut off from urban life because the Asian population group blocked their entry to trade and commerce while participation in the administration was denied them by the Europeans. However, since Kenya received her independence and a moderate Kenyanization is being carried out, a fundamental structural change is taking place among the urban population. Many Africans have pressed into the towns, and many have taken over trading and commercial firms or moved into positions previously held by Europeans and Asians in the administration, schools and health services. In effect this process might well be described as the Kenyanisation of the towns of Kenya. But neither the towns' nor the country's economy were equipped to cope with the influx of Africans into these towns. In Nairobi and Mombasa the formation of slums could not be prevented. As Table 5 shows, Kenyan towns experienced a considerable rate of increase among the African popula-

Table 5. Urban population Kenya

Size class: Inhab.: Number of places:	2,000 – 4,999 26	5,000 – 9,999 12	10,000 – 99,000 11	over 100,000 2	over 2,000 total 51	Urbanisation in the population group
Africans	78,126	65,791	145,883	608,226	898,026	8.4
Europeans	416	406	3,205	24,110	28,137	69.3
Asians	2,909	2,725	18,872	106,238	130,744	94.0
Arabs	416	2,449	2,218	16,934	22,017	78.6
Other	19	25	89	851	984	49.5
Total	81,886	71,396	170,267	756,359	1,079,908	9.9

Source: Kenya Population Census, 1969, Vol. II, p. 28 et seq.

tion between the census of 1948 and that of 1969. Nevertheless, the degree of urbanisation among the African population—8.4 per cent in 1969—was the lowest urbanisation rate of all the Kenyan population groups. Table 5 shows that 94 per cent of the Asian, 78.6 per cent of the Arab and 69.3 per cent of the European population in Kenya live in towns of over 2,000 people.

4. Health Problems in the Urban Sphere

a) National health: Housing, water supply, refuse removal and sewage disposal are the three most important material problems in the urbanisation process which have a direct relation to health. Migration, social disintegration, disrupted sex ratio, unemployment, poverty, criminality, promiscuity, and alcoholism are the human problems of this process. Both complexes together determine the pattern of diseases in tropical urban centres undergoing rapid population growth, whereas purely tropical diseases fade into the background.

The adaptation of a rural population to the conditions of life in the city contains special problems. Whereas in horizontal slum areas a few square yards of earth, a banana plant, and a few chickens may act as a reminder of old, familiar surroundings, life in a multistorey block, though hygienically more advantageous, may not offer adequate satisfaction to people. The problem of slums transfers itself to the vertical dimension.

Housing programmes result in expenses for prospective occupants and presuppose an income adequate to meet the payment of rent. So far only Nairobi and Mombasa have experienced these world-wide problems to any marked degree, whereas Nakuru, Kisumu, Eldoret, Thika, and Kitale follow in their footsteps so far as extent, if not the growth rate, of the problem is concerned.

Almost 8,000 dwelling units are required annually in Nairobi and Mombasa alone; this is 80% of the housing space required in towns on account of population growth (see Table 6). The requirements of slum clearance are not included in these figures. In 1972 somewhat less than half of the Nairobi households enjoyed solid, permanent dwellings with water supply and sewage system. Whether legally or illegally, the other half occupied shanties or mud huts without water and sanitary facilities within acceptable reach (Urban Study, Nairobi City Council, cited by Mugo, 1974).

The National Housing Cooperation, which makes funds and technical assistance available to local authorities, and where necessary directly undertakes housing projects, will be unable to solve this problem in the foreseeable future. In 1972, 2,497 units were completed, 60% of which were in Nairobi and 20% in Mombasa (Economic Survey, Nairobi, 1973, p. 197).

The immense problems of *water supply* in the rural areas of Kenya are aggravated by great difficulties in the case of urban requirements. Population growth, as well as increasing industrialisation, contribute to it. But whereas the budget for the rural areas rose twelve-fold from K£ 152,000 in 1968/69 to K£ 1,880,000 in 1973/74,

Table 6. Urban population growth and housing requirement

Urban 1 Centres	Population estimate 1974	Annual increase %	No. of Households 1974 (per 5.15 pers.)	Housing units required aver./ year	% of total
Nairobi	668,000	5.7	131,400	5,880	58.0
Mombasa	305,000	4.5	60,000	2,000	19.7
Nakuru	74,500	5.7	14,600	640	6.3
Kigumu	45,200	5.6	8,900	400	3.9
Eldoret	41,200	6.4	8,100	420	4.1
Thika	37,200	8.5	7,300	500	4.9
Kitale	13,000	2.8	2,600	60	0.6
Subtotal	1,184,100	5.4	232,900	9,900	97.5
Urban area council	117,100	3.0	22,700	360	2.5
Total Urban areas	1,301,200	5.2	255,600	10,260	100

Source: Development Plan 1970 – 1974, p. 509

the rises for urban areas were 2.4 times as great, that is from K£ 150,000 to K£ 360,000 between 1968/69 and 1973/74 (Development Plan 1970 – 1974, p. 376, Nairobi 1969). On the other hand, it must be remembered that only some 10% of the population lives in urban areas.

The third problem is *sanitation*. In 1965 it was estimated that only 49% of the African urban households had use of a water closet, and even most of these were shared (Nairobi 72%, Mombasa 22%). Pit and bucket latrines ranged from 65% in Mombasa to 24% in Nairobi. Nairobi City Council has replaced bucket latrines by water closets (Mugo, 1974).

Development expenditure in what is for urban communities an important field are still very low. The municipalities' budget for this amounts to K£ 600,000 – K£ 700,000 annually (Development Plan 1970 – 1974, p. 376, 1969).

An exact survey of the health problems of the two largest cities cannot be provided since reporting is deficient and diagnoses in the smaller health units are unreliable. The most important diseases of childhood are respiratory and gastro-intestinal ones, together with bronchial pneumonia and measles in conjunction with malnutrition. Malaria does occur in Nairobi and Mombasa, but transmission does not occur there to any marked degree. Many cases of malaria are imported from elsewhere. *Anopheles gambiae,* the vector, is relatively demanding concerning the water purity of its breeding places. Pollution of water-courses in urban districts is therefore a deterrent to the development of malaria vectors. The vector of filariasis prefers the polluted waters of Mombasa. The health authorities attempt to keep the breeding places under control by spraying.

Schistosomiasis is endemic in Nairobi. In some school classes 20 – 70% of the children were found to be infected. Rivers and streams in Nairobi are infested by the transmitting snails, whereas the mainly brackish water in the immediate urban district of Mombasa is unfavourable for the development of snails.

Malnutrition is less a problem of general supplies than one of individual buying power. More than half the urban population earns less than K£ 20 a month. Nutritional habits undergo modernization, which is decidedly not for the better. In particular, the propagation of artificial baby foods in the urban context is the cause of considerable problems.

The municipalities of Nairobi and Mombasa, as well as the remaining municipalities, still run their own health services, which were not taken over by the central government in 1970, as was the case with the other County Council Health Services. The Mombasa and Nairobi health services are particularly burdened by the large and unknown number of patients from outside the jurisdiction of the municipalities who present themselves for treatment, which effectively makes inoperative the comparatively dense network of services in urban as opposed to those of rural areas (see Chapter B. 5).

The environmental health staff, the health inspectors and the sanitary inspectors fulfill an important role by means of their control hygiene in markets, the grocery trade, and eating houses, as well as by rat control, water supplies, sewage disposal, and the industrial and school health services. Another important task is the check made on domestic Aedes mosquitoes, the potential transmitters of yellow fever, and the destruction of detected breeding places (City Council of Nairobi, Annual Report of the Medical Officer of Health, 1965, 1975; Mombasa Municipality Medical Officer of Health, Annual Report 1970).

b) International health and port health in Nairobi and Mombasa: Special tasks fall to the Airport and Port Health Authorities in Nairobi and Mombasa owing to international sanitary regulations and the Public Health Act. As regards the export and import of contagious diseases required to undergo quarantine, their proficiency is at least documented by the fact that there were no such importations via seaports or airports over recent years.

IV. Fundamentals of Economic Geography

Kenya is an agricultural country, poor in natural resources, in which more than four-fifths of the population are compelled to live directly or indirectly from agriculture. Nevertheless, Kenya belongs to the group of most highly developed countries in tropical Africa. The British influence through the colonial period has left its mark more deeply than in most other countries. In particular the establishment of the White Highlands exercised a lasting influence on the economic development. Although it cannot be denied that the development of the White Highlands was to the disadvantage of the African population groups and that in some areas com-

peting claims on land were made by European farmers, African smallholders, and herdsmen, the establishment of the White Highlands has nevertheless greatly accelerated economic development, especially hastened the completion of the infrastructure, and enabled Kenya to proceed with the expansion of its economy after gaining independence. Over a period of one generation, Kenya developed from a simple African subsistence economy to a dual economy in which, at the present time, the value of production outside the monetary economy amounts to only about one-fifth of the gross national product. Within the monetary economy agriculture produces about

Table 7. Employment by Industry, 1974

Agriculture and Forestry	261,148
Mining and Quarrying	3,869
Manufacturing	101,332
Electricity and Water	5,694
Construction	44,437
Wholesale and Retail Trade, Restaurants and Hotels	57,021
Transport and Communication	46,310
Financing, Insurance, Real Estate and Business Services	21,896
Community, Social and Personal Services	284,556
Total	826,263

Source: Statistical Abstract 1975, Table 221

one-fourth of the total value of the economy, followed by manufacturing and repairing. Due to the lack of natural resources, the Kenyan economy is concentrated above all on the processing of agricultural products produced in the home country. Since independence in 1963, considerable economic growth can be observed: in the period between 1964 and 1967 an average annual rate of increase of 10 per cent for all enterprises (at constant 1964 prices) has been accepted for the monetary economy (see Economic Survey 1973, Table 1.2). But the rapid growth of the economy does not veil the fact that the living conditions among the farming population in Kenya's densely settled areas have deteriorated due to the increased pressure for land that has forced more people to leave farming than can find employment in other branches of the economy.

In 1971 Kenya had around 11.7 million inhabitants. The monetary economy, however, employed only about 700,000 persons. The number of those people of working age who are underemployed or unemployed remains unknown.

1. Agriculture

In Kenya agriculture presents numerous modes of land use. The mode of agricultural land use is the synthesis of the farming systems, agricultural methods, systems of land tenure, and other institutions and techniques which are used by the farming population engaged in agricultural production. Ecological, socio-economic and political conditions are dependent factors. In this synthesis with its cartographic representation of the facts, the following are considered:

(a) cultivated crops, types and their relative proportion
(b) livestock, types and their relative proportion
(c) the role of animal rearing on farms, or rather in the farm budgets
(d) agricultural methods (cultivation by hoe and plough, use of machinery)
(e) the system of land ownership and the structure of farm size
(f) the farms in relation to the market (Hecklau 1970, p. 3).

The great diversity of styles of agricultural land use in Kenya is the outcome of the activities of very varied population groups, which, as already noted, can look back to most varied social, racial, and cultural origins. These diverse population groups have encountered extraordinarily diverse ecological conditions in their turn, as has been shown in the section on the basic physical geography. The altitudinal differentiation of the country and its location on both sides of the equator have resulted in such diversified environmental conditions that products of the temperate climates like wheat, barley, potatoes, and vegetables can be grown not far from tropical crops like bananas and coffee.

Two main categories of land use can be distinguished: the traditional styles of land use of the Africans, and those introduced by the Europeans in the former White Highlands. The forms of land utilisation initiated by the Africans are characterised by progressive differentiation. Almost all stages of development are to be found, from the simple subsistence pastoral economy to the modern rotational grazing, from primitive shifting cultivation to farm holdings which are run in accordance with the findings of modern tropical agricultural science.

Land use styles introduced by the Europeans can be classified under three basic forms:

a) Ranching
b) Mixed farming
c) Plantation economy

Styles of agricultural land utilisation, dependent on socio-economic and natural ecological conditions, are the basis for the economic and social areal differentiation of Kenya presented in Map 5 and described in Chapter D of this study.

a) Small-scale agriculture and cattle keeping: The common socio-economic characteristic of peasant farming and cattle herding societies is their existence at the level of a subsistence economy, their production being intended only in part for the market. The peasant farming families seek to provide for their own food requirements from the produce of their own holding. Surplus produce is sold. If energy and land are available in addition to this the peasant farming family will grow additional crops intended specifically for cash sale. Cattle keeping peoples live from their stock, selling or exchanging it only in order to obtain vegetable foodstuffs or consumer goods for their own needs, or in order to pay taxes and similar imposts to the state.

In most parts of Kenya maize is the staple food of the peasant population and thus the most important crop. The grains are pounded into flour, which is called posho, and then cooked into a thick porridge which may be supplemented by vegetables, meat, or other pulse crops (Photo 32).

Thanks to advances in plant breeding, in which Kenya's own agricultural experimental stations have played a substantial role, hybrid maize provides comparatively high yields. The so-called katumani maize ripens within three months, and can therefore be successfully cultivated on the margins of rain-fed agricultural zones.

In many areas maize has largely replaced the indigenous types of millet, in spite of their higher nutritional value, since maize provides higher yields and is more easily harvested and stored. In western Kenya, and particularly in the coastal areas of Lake Victoria, a great deal of sorghum is still planted where ecological conditions are unfavourable or the natural productivity of the soil has been exhausted by overcultivation.

Before the coming of the British two tuber cash crops from South America, sweet potato and cassava, were already widely known in East Africa. The colonial power propagated their cultivation as reserve crops in the event of famines caused by poor harvests after drought. Cassava will thrive even in nutrient-poor or exhausted soils and with little precipitation; in addition it can be left for up to four years in the ground without perishing. Cassava is frequently grown as the last crop in the cultivation cycle. Because of its high yield per acre cassava cultivation enjoys greater popularity among the peasant farmers than is desirable for the health of the population. Unbalanced feeding on cassava, which consists almost solely of carbohydrates, leads to the well-known protein deficiency disease, kwashiorkor. Cassava is widespread in the coastal area and in western Kenya at altitudes below 1,500 m (Photo 21).

Bananas, which were introduced from Asia, have been grown in East Africa for centuries. Cultivation conditions in Kenya are not as favourable as in neighbouring Uganda, but at higher altitudes where they receive more precipitation they can be seen as small clumps on many smallholdings.

Various legumes, particularly beans, are of great importance to the nutrition of the peasant population in almost all farming districts. Rice, groundnuts, sesame, onions, vegetables, and other crops are not cultivated everywhere, and in any case occupy only very small plots.

Up to the fifties the Kenyan peasant farming population grew only subsistence crops; those surpluses which accrued if the harvest was good were then sold.

Up to the 1950's the colonial administration did little for the development of African peasant agriculture. In the beginning of the colonial period the development of the White Highlands was given priority. From 1920 onwards attempts were made by the colonial administration to restrict the paramountcy of white interest and to follow a dual policy of economic advancement of all the population groups in Kenya (see, for example, B. B. Dilley, 1966, p. 279 et seq). The results were extremely modest. On the contrary, a great increase in population resulted in considerable land scarcity in the densely populated native reserves. Equal division between heirs, or "Gavelkind" tenure (undivided inheritance), reduced many peasant holdings to a size below the minimum able to sustain a family. Repeated sub-division rendered cultivation difficult. The traditional management led to soil erosion and thus impaired the natural conditions of production.

The agricultural development policy in the later years of the colonial administration as set out in the "Plan to Intensify the Development of African Agriculture in Kenya"— shortened to "The Swynnerton Plan" after its initiator—enabled the smallholders to grow cash crops like coffee, tea, sisal, pyrethrum, wheat, rice, vegetables, fruit, potatoes, and others. In spite of all the problems, the development proved so successful that African peasant holdings now grow half the products marketed in the country's agricultural sector.

Masefield (1962), Ruthenberg (1966), and L. H. Brown (1968) have described the fundamental structural changes in peasant agriculture, which were begun by the colonial administration and purposefully continued by the Kenyan government. These led to the commercialisation of African agriculture and had great social effects, which will be outlined later.

aa) Livestock: By comparison with agriculture, animal husbandry plays a subordinate role with most African peasants. Livestock—Zebu cattle, goats, sheep, and chickens—are not integrated with farm management as has been the rule in European farming until most recently. Cattle, goats, and sheep are driven out to fallow land or into the bush to graze. The tractive power of animals is only employed in a few areas and the manure is used only by progressive farmers; in the reverse direction, where possible mutual benefit might occur, no provisions such as the growing of special fodder for the animals are made, and as a rule, milk and meat yields are very low. Since the end of the colonial period many Kenyan farmers at altitudes higher than 1,500 m have replaced their less valuable Zebu stock by European dairy cattle. In the process of agricultural intensification, they have laid out seeded pastures and feed and look after the dairy animals. As an illustration, this requires a twice-weekly spraying with insecticides against disease-carrying ticks. The sale of milk has become one of their most important sources of income.

Any further improvement in animal husbandry might well constitute a fundamental improvement in the nutritional situation of the population, especially as regards protein supply.

It is regretable that livestock is far too highly regarded as a status symbol or as a reserve against needy times. What counts here is the number of heads, rather than

their age or productive power, and in consequence herds have frequently grown too old. This situation applies particularly to the animals of the nomadic herdsmen, and it will be discussed in detail in Chapter D on "African pastoral societies in predominantly arid areas".

bb) Forms of agricultural techniques: Hoe cultivation is the most widely employed agricultural technique among the peasant farming population of Kenya. The smallness of their holdings, often situated on the slopes in the Kenyan highlands, together with the lack of pastures to feed draught animals, combine to impede the spread of the plough.

A tradition of plough cultivation has been established for several decades in some areas of western Kenya, especially among some erstwhile herding tribes like the Kipsigis, Nandi and Luo.

In most areas there is a tractor hire service, and the farmers can make use of it when ploughing their fields. The introduction of plough cultivation, particularly in connection with the tractor hire service, and the employment of paid labour, has enabled the farming families to cultivate much larger acreages than hitherto. Together with the cultivation of marketable crops, it leads to a marked increase in the value of land.

In many areas, chiefly those in densely settled upland location, serious damage has been done by erosion. In its fight against erosion, the administration has ordered the construction of many thousands of kilometres of field terraces, protective grass strips, and even stone walls by the Africans. Such works have often been undertaken communally.

In Kenya population growth has led to a change in the form of farming systems. Shifting cultivation was commonly practiced in pre-colonial times. In the interim farmers in many parts of the country have gone over to rotation of crops, with semi-permanent arable farming. As land became scarcer, densely settled areas finally accepted a predominance of permanent arable husbandry, in which the number of cropping years of any given field exceeds the number of fallow ones. The necessity to shorten the number of fallow periods more and more confronts the farmers with the problem of declining fertility and yields from their land. In former times farmers already attempted, albeit unconsciously, the optimal utilisation of the nutrient cycle by crop rotation. However, a start in the fundamental alteration of cultivation habits among the African farmers was made only in connection with the individualisation of land holding and planned management based on scientific agriculture, which is applied by progressive farmers. Farm planners employed by the administration draw up plans for the farmer which are exactly adapted to the conditions of the farm in question. Rotation adapted to the ecological conditions is recommended, and thus the regenerating bush fallow is replaced by the laying out of grass fodder areas linked with the keeping of high-value breeds of dairy cattle. To date, this procedure has been carried out only among a few, so-called progressive farmers in the higher parts of Kenya.

cc) Land tenure: The basic principle of African land tenure can be summed up as follows: everyone who cultivates land also enjoys the right of usage of that land, which in most cases he will be able to leave to his heirs. Property that has been allowed to revert to its natural state, however, will revert to the community. Uncultivated, communally owned land, can be used by all members of the community as cattle pasture. The headman of a clan, or in frequent cases a council of elders, is entitled to transfer some of this communally owned land to an individual for his own use. In pre-colonial and early colonial times only a few densely settled agricultural areas had evolved a de facto land ownership as a result of land scarcity. Almost throughout Kenya the process of population increase has, however, resulted in the development of the facto land ownership without state intervention, insofar as families have ceased to practice shifting cultivation and no longer permit cultivated land to revert to the wild state. Land ownership which has developed in this manner is usually recognized by the neighbours as such without any formalities. However, the scarcer the land becomes, the more often do boundary or demarcation disputes arise. Security of individual land title is the pre-condition for the safeguarding of credit in the form of mortgages. Credit, in its turn, is indispensable for development of peasant economies, and land is the sole collateral that the smallholder can offer as surety for such credit.

A start has therefore been made in Kenya with the surveying, demarcating and recording in the Kenyan Land Register of the peasant farming areas. This development has gone furthest in the areas of the Kikuyu, Meru, and Embu tribes. Together with this agrarian reform, the fragmentation of land ownership has been overcome. The individualisation of land ownership and the commercialisation of African farming has led to a greater differentiation of hitherto more egalitarian peasant societies. Successful farmers can enlarge their holdings by employing paid labour, applying modern forms of agricultural technology, and growing cash crops on a larger scale. Less successful farmers cannot enlarge their acreage. The customary mode of gavelkind inheritance further reduces the size of their holdings to the point when the available land has become too small to support the peasant family. This social differentiation is leading to more and more peasants being forced either to leave the rural areas, or to seek employment as labourers for more successful farmers.

dd) Relation of peasant farms to marketing: As has already been described, the paramount aim of the peasant population is to cater to its own nutritional requirements, and to produce for the market only if either a surplus has arisen or the family owns sufficient land and labour for exclusive production for the market. The degree of commercialisation of peasant holdings varies greatly. In the vicinity of urban markets more is produced for sale than in remote areas.

Agricultural products which predominantly serve the domestic requirements of the peasant population can be traded in local markets where no statistical record is kept.

"Every year one million small-scale farmers in Kenya grow maize worth nearly 25 million pounds, of which they barter or eat 95 per cent. The remainder of the crop goes to market . . ." (Who controls industry in Kenya, 1968, p. 1).

The marketing of crops grown for sale, especially those destined for export, is carried out or supervised by cooperative organisations, by statutory corporations, or by the authorities.

A summary of crops cultivated by African small-scale farmers during the rains is given in Table II (Annex).

b) Large-scale ranching, mixed farming, and plantations: Traditional African styles of land use differ in all respects from those of what are nowadays called "the large farm areas", which include the former White Highlands as well as plantations on the Kavirondo Gulf, in the Voi and Taveta areas, and on the coast of the Indian Ocean.

Several population groups can be regarded as the initiators of this form of land utilisation: namely gentleman farmers from Great Britain with their overseas properties, Boers with long-established agricultural traditions who immigrated from South Africa, as well as some entrepreneurs who originated in the Indian subcontinent. These Asian entrepreneurs have been the chief founders of the sugar cane plantations and the sugar industry in East Africa.

Essentially these population groups only carried out management activities, for the elementary manual labour was done by Africans.

The decision taken by the British to develop Kenya on the basis of large-scale farming is usually attributed to the claim that the cost of constructing the Uganda Railway from the Indian Ocean to Lake Victoria had to be redeemed. The building of the railway was carried out for political, strategic, and economic reasons. It was intended to safeguard the occupation of East Africa by Britain and to link the economically further advanced Kingdom of Buganda with world trade. In Kenya, however, the railway ran across sparsely settled and economically underdeveloped country.

Ernst Weigt (1932), Elspeth Huxley (1953, 1957, 1962), W. T. W. Morgan (1963), R. S. Odingo (1969, 1971) and others have discussed the creation, the development, and the present economic problems of the former so-called White Highlands.

In contrast to most African countries, the forms of land utilisation initiated by the Europeans in Kenya, namely large-scale farming, ranching, and plantations achieved a dominant position in the economic life of the country. Shortly before the independence of Kenya the large-scale farms produced about four-fifths of all agricultural exports on one-fifth of the agriculturally-utilisable land.

Areas poor in precipitation, that is with less than around 760 mm precipitation, practise extensive pastoralism to produce beef cattle. Wherever the ecological conditions are more favourable for the pastoral economy and the location in relation to marketing and processing centres likewise fortunate, this is supplemented by the keeping of dairy cattle.

Those areas somewhat better off in their precipitation, that is with more than 760 mm, practise mixed farming. At lower altitudes in the Large Farm Areas maize predominates, while wheat is the chief crop grown at the higher altitudes, supplemented in some areas by malting barley. Pyrethrum and coffee are special crops cultivated on many farms. The arable acreage amounts only to approximately one quarter to one-third of the total farm holdings, the remainder being left fallow and serving as natural pasture for the extensive ranching. Cereal farming is supplemented by cattle rearing and here and there with sheep to a modest extent. Dairy cattle only came to enjoy greater importance after the second World War.

The development of mixed farming has been greatly hampered by a lack of capital, bad marketing conditions, remoteness from the world market, and by political events during this century. Although already before the outbreak of the second World War, the fact had been accepted that nothing but a well-ordered ley farming in conjunction with a balanced rotation of grain crops and pasture cultivation together with animal husbandry would guarantee an optimal utilisation and maintenance of the productive capacity of the soils, farmers were able only some years after the war to proceed to put their knowledge into practice.

Plantations: Ruthenberg (1967, p. 183) defines plantations as those large scale farming enterprises of the tropics which include the industrial processing of their produce within the unit. On Kenyan plantations sisal, sugar cane, tea, and coffee are grown, and to a modest extent even coconut palms on the coast; this will be described in Chapter D. Table III (Annex) shows the land utilisation in the Large Farm Areas.

2. Mining and Quarrying

Although numerous minerals have been discovered in Kenya, most deposits are too small to render their mining on a large scale a profitable proposition. Success has so far eluded the search for coal and oil. The mining of magnetite (9,240 tons valued at £ 57,450) began in 1972, whereas gold mining, which recorded a profit of £ 448,000 in 1968, is no longer of significance. The output of fluorspar ore (fluorite) in the Kerio Valley has shown a considerable increase from 192 tons in 1968 to 10,457 tons in 1972. A new mining development—for zinc, lead and silver—is in progress on the coast near Kinangoni.

In 1972 only the output of potash, salt, limestone, and fluorite exceeded an annual value of £ 100,000. The largest and economically the most important deposits are found in Lake Magadi, where potash and salt are mined by the Magadi Soda Company Ltd., a subsidiary of Imperial Chemical Industries (I.C.I.).

Cement factories in Kenya are situated north of Mombasa, where coral limestone is used for the production of cement, and near Nairobi on the Athi River.

Table 8. Mineral output, 1972

	Quantity (metric t.)	Value (Kenya £)
Soda Ash	161,260	1,889,226
Salt (refined)	43,406	424,785
Limestone products	28,127	232,486
Fluorspar ore (Fluorite)	38,226	6561

Source: Economic Survey, Table 5.12 and 5.13.

3. Manufacturing and Repair

Development of this sector on a larger scale began significantly only after World War II. In 1971 only about 90,000, i.e., 13 per cent of the Kenyan work force, were employed on this side of the economy. Most firms involved are very small, and many are not even recorded in the statistics. But according to the 1971 "Survey of Industrial Production" (Table 3 of this survey), 291 firms with more than 50 employees each employed 76,366 workers, and together accounted for about 80 per cent of the entire output. Kenya has no class of craftsmen endowed with the skills which might have become the carriers of industrial development. Small and medium-sized firms have in the main been developed by members of the immigrant Asian community, who, as traders, have accumulated capital and invested it in industrial enterprises. In the larger firms there is a dominance of British capital, and in 1968 only four of Kenya's top fifty directors had Kenyan citizenship. Although Kenyanisation of the economy is the aim, "with few exceptions, Kenyans are tending to enter the modern economy at the higher levels as merchants or as landowners" (Who controls industry in Kenya, 1968, p. 261).

By far the largest part of the gross national product of the economic sector "Manufacturing and Repair" stems from the processing and fabrication of agricultural raw materials (1971=29%), above all from the production of foodstuffs and luxury goods based on indigenous raw materials. Examples of these include meat and milk products, vegetable and fruit conserves, grain mill products, bakery products, sugar and sugar confectionary, spirits, beer and soft drinks. The agricultural non-food processing and fabricating industry also plays a certain role in the manufacture of textiles, leather goods, vegetable oils, tobacco products, pyrethrum extract, and others.

Almost 20,000 persons, that is a quarter of all those employed in the manufacturing and repair sector, are concerned with the maintenance and repair of transport. Of these 10,725 alone are concerned with the railways.

Kenya's industrialisation encounters considerable difficulties, for there is an abundance of unskilled labour matched by a lack of indigenous specialists. Foreigners, of course, demand higher pay. Bearing in mind the low living standards, it is difficult to achieve any capital formation among the indigenous population. Foreigners will only invest in Kenya if the interests rates on their capital are high. But the greatest obstacle to rapid industrialisation is, however, the limited capacity of the East African market for absorbing the goods offered. Kenya, Uganda, and Tanzania form the East African Customs Area, which is about seven times the area of the Federal Republic of Germany yet inhabited by scarcely over half its population. Moreover, the purchasing power of the African population is very low, so that the demand for goods is too small to justify their production in the country. "The market for cement and textiles is large enough to make local production economical but that for motor vehicles is not" (O'Connor 1971, p. 156).

4. Energy Supply

Kenya's energy supply is founded on a very narrow base. Oil and coal deposits have so far not been discovered. Seven international companies control the importation of crude oil, which is refined in Mombasa. Oil products, chiefly petrol, are sold not only in Kenya, but also in the neighbouring countries of East Africa.

Hydro-electric power stations on the Tana River and Mt. Kenya and thermal power stations in Nairobi and Mombasa meet about two-thirds of the Kenyan electricity demand. About one-third is imported from the hydro-electric power station at Jinja on the Victoria Nile in Uganda. The backbone of Kenya's distribution network is the 132 Kw. grid, which leads from Uganda via Nairobi to Mombasa. It guarantees alternative supplies to the grid through predominantly hydro-electric power during the rainy season and more thermally-generated power during the dry season. Away from this grid the more isolated urban areas have their own diesel stations. All towns in Kenya therefore have the benefit of electricity, and Kenyan development policy aims at supplying the densely settled rural areas with electricity as well. Nevertheless, by far the greatest part of the rural population will have to make do without domestic electricity, since the connection of the widely dispersed hutted settlements with the main grid is not an economical proposition. Demand for electricity has shown an annual rate of increase of some 10 per cent, and for this reason further hydro-electric power stations are to be built on the Tana River. All of Kenya's other rivers are too small for the production of electricity on a large scale.

Investigations are under way to find out the ways in which geothermal energy in the Rift Valley might be exploited (Development Plan 1974 – 1978, Part 1, p. 313 et seq.).

5. Forestry

The natural vegetation of the Higlands of Kenya up to the tree-line, at an altitude of about 3,000 m above sea-level, consists of different forms of dense, tall, evergreen forest with a wide variety of species. By far the greater part of this primeval forest has by now been cut down, and only about one-sixth of the region with more than 850 mm mean annual precipitation has kept its forest cover.

The forest covers those regions of the country which are ecologically not favourable for agriculture and human settlement. Already in pre-colonial times large tracts of forest had been cleared and put under agricultural usage by the peasant population. Population growth led, as described earlier, to an increasing shortage of land, and larger and larger tracts of forest fell victim to clearance. "The fact that the land best suited to the growth of forests is also good agricultural land presents a serious and continuing problem for forest management policy. As the rural population continues to increase so do the pressures of converting forest lands to farms. The catchment protection role of forest lands has been seriously compromised in recent years. Illegal, spontaneous settlement, with continuous maize cropping, has not only removed the forest cover but led to greatly accelerated runoff and soil erosion over large areas. Removal of the natural forest cover has also led to reduced and erratic stream flows in many places" (Development Plan 1974 – 1978, Part I, p. 259 et seq.).

Large, compact, natural forest areas are represented by the Mau Forest, the montane forests in the Aberdare Range, on Mount Kenya and Mount Elgon and in the Cherangani Hills—to name only the most important ones. Like all African primeval forests, these, with their great specific variety, contain only few economically profitable species which are characterised by very slow growth. The transformation of the natural forests into plantations in the course of Kenya's economic development cannot be avoided. As a rule it is carried out as clear-felling, followed by afforestation with cypress and pine, although often with a short period of agricultural utilisation. Another approach consists of felling the wood trees and afforestation of the gaps with tropical hardwoods. The value of the natural forests can be enhanced in this way.

About 1.7 million ha of land have been gazetted as forest areas by the government, but only little over half these forest areas corresponds to European conceptions of forest. The gazetted forest areas (excluding private forest land) are closed forest (Hectares 940,000), woodland (Hectares 336,000), bamboo (Hectares 151,000), grassland (Hectares 211,000), and mangroves (Hectares 45,000) (Statistical Abstract 1975, Table 97).

All these vegetation formations in the gazetted forest areas have important functions to fulfill "concern for the maintenance of forest lands is not confined to considerations of supply and demand for timber. It would be difficult to justify the present areas of forests, let alone their expansion on this basis alone. The more fundamental contribution of the forest areas is the essential role they play in protecting and enhancing the surface and ground water supply, and in controlling soil erosion.

Over 1.3 million hectares of the national forest land have been designated as catchment protection forests" (Development Plan 1974 – 1978, part I, p. 259).

Closed forests correspond to the natural forests described above; only about 100,000 ha are plantations with exotic softwoods, chiefly consisting of cypress and pine. However, these plantations supply more than half the production of softwood timber. Both modern forestry and the timber processing industry are being purposefully developed. "Forest-related industries account for about 6 per cent of all non-agricultural employment. Approximately 30,000 persons are employed in forest exploitation and the manufacture of wood and timber products. Several major forest development programmes are directly related to the raw material requirements of these industries. Such programmes include forest plantations, the Sawmill Extension Service and the construction of access roads for the removal of felled timber" (Development Plan 1974 – 1978, Part I, p. 261).

Much more important though than modern timber production is the supply of fuel and construction timber to the African population. "Total domestic consumption of forest products was estimated to be 12 million cubic metres of roundlog equivalent in 1970. Approximately 92 per cent of this consumption was for fuelwood; only 5 per cent for industrial purposes — sawn timbers, pulp and paper and plywood. Wood will continue to be the principal fuel in rural areas in the foreseeable future, and the demand for wood for this purpose and for building poles is expected to grow at a slightly higher rate than population, i.e. about 4 per cent per year" (Development Plan 1974 – 1978, Part I, p. 258).

The population not only takes fuel from the closed forests but also from the natural bush outside the gazetted forests. Wherever possible farmers have been supplied with their own piece of land for the purpose of fuel production. Wood is not only used for cooking but also for heating in the cool highlands. In the future it will not always remain possible to supply the growing population with sufficient wood as fuel, and the development plan has already put on record the necessity of providing alternative fuels.

Large quantities of wood are burned to charcoal, great amounts of which are sold in the towns, although demand has decreased (O'Connor 1971, p. 142). For the farming population the supply of wood for fuel is of an importance equal to that of building timber, since their houses are constructed as a framework of poles (see Photo 18), the walls of which are subsequently decked with loam and the roof with grass or corrugated iron (as described in Chapter III.1).

6. Fisheries

The fishing industry ought to contribute significantly to the protein supply of the population, particularly the poorer sections of the population in towns and densely settled higher parts of the country in which kwashiorkor is frequent and there is a need for higher protein

levels than is the case with present nutrition. Consumption of fish varies greatly among the different groups of population. While it is a traditional part of the diet among those in the vicinity of lakes and rivers well stocked with fish, as well as on the Indian Ocean coast, farming and herding families away from such waters consume little or none. The average consumption of fish among the Kenyan population, 3 kg per person annually, is extremely low if compared with the Federal Republic of Germany, for example, in which the yearly average exceeds 10 kg per head.

The extent of water surface within the territory of Kenya is about 7,000 km², added to which there is the share of the 63,000 km² Lake Victoria and about 1,000 km of coastline bordering the Indian Ocean. Until recently some of the lakes of the interior were regarded as being well stocked with fish, but at present the catch is either static or even declining, as shown in Table IV (Annex).

In some cases, as in Lake Naivasha and Lake Victoria, the decline in catch is to be attributed to selectively excessive fishing; in others, ecological problems seem also to contribute — such as heavy silting in Lake Baringo and the expansion of salvinia in Lake Naivasha (Development Plan 1974 – 1978, Part I, p. 266).

Lake Rudolph, Kenya's largest lake, is very rich in fish stocks, but the development of a fishing industry is hampered by the difficulties of the lake's remote location in the north of the country in an area which is semi-arid, sparsely populated, and little opened-up by transport. No all-weather road so far leads from Lake Rudolph to densely populated areas with potential markets which could permit normal transport by lorry. The distance of the planned road from Lake Rudolph to Kitale, where the greater settlement density begins, alone amounts to about 500 km.

Fishing is almost solely carried out by African fishermen, who are equipped with small boats and simple fishing gear. Only a few boats are equipped with outboard motors. An attempt has been made to organise the fishermen in cooperative societies and to equip them with modern craft and better gear, but the results are scarcely encouraging.

Fishing in the Indian Ocean has so far been limited to inshore fishing with small dugout canoes. Deep-sea fishing requires high investments, the economic viability of which is still being investigated. The expansion of inshore fishing is limited by the take-up of the market among the coastal population. The distance to the densely populated highlands is great and transport routes are limited to the 500 km-long main road and the Mombasa-Nairobi railway line.

The fish trade is in the hands of small dealers. Fish can only be sold fresh in the vicinity of the fishing grounds. By far the greater amount is smoked, salted, or dried in the sun. The fish species of the inland waters which are of greatest importance are Tilapia and Haplochromis.

The Kenyan Development Plan envisages the modernisation of the fisheries, fish processing, and fish marketing. A special tourist attraction and a leisure activity for Kenyan residents is the sport of fishing on the inland lakes as well as the Indian Ocean.

7. Internal and External Trade

In Kenya, as elsewhere in the world, the origins of trade are to be found in barter. It was small in extent because the population lived at the level of a subsistence economy. Almost all that the family required in the way of food, clothing, tools, and huts for dwelling or storage, was produced at home. But surplus products were exchanged for commodities, such as pottery and iron tools, which could not—or at least not easily—be made on the spot. So too, foodstuffs were exchanged, and in particular the products of the field were exchanged for those of animal origin by farmers and herdsmen.

New forms of trade developed with the gradual transition from barter economy to money economy. The establishment of the colonial administration in Kenya was accompanied by the advance of traders—predominantly members of the Asian community—all over the country and into the most remote of the regions. These people opened dukas (rural stores offering a variety of goods) in the so-called trading centres, which were normally located in the vicinity of administrative or police stations. In these dukas the traders offered commodities to the farming population, which ranged from textiles and simple domestic items to provisions such as salt, tea, and spices, and much more besides. Those intent on acquiring such goods were now obliged to grow more agricultural produce than was needed for their own requirements. The Asian traders operated as middlemen, who took the farmers' surplus produce and provided other commodities. Foodstuffs like maize and millet were sold to the growing sector of non-agricultural population, while other agricultural products were collected for export. In this way the Asian traders contributed considerably to the commercialisation of African smallholder agriculture. The Africans were unable to assert themselves against the competition of the Asian traders, and until recently retail as well as wholesale trade were thus almost exclusively in the hands of the non-African and predominantly Asian traders. When, during the course of the Swynnerton Plan's implementation, the government initiated the cultivation of export crops by the Africans, state or para-statal marketing organisations were set up in order to arrange the sale of crops intended for export such as coffee or tea. But the trade in foodstuffs offered by farmers is still conducted by private traders insofar as the farmers prefer not to sell their surplus produce in the market themselves. Until recently these traders were mostly members of the Asian community, but as a result of the policy of Africanisation they have now largely been replaced by Africans.

"Most small commercial firms have already transferred to citizen ownership. Larger and more intricate firms still remain in the hands of non-citizens. The Government will employ all necessary measures, including training, expansion of credit, and trade licensing to complete the Kenyanization of commerce by the end of the Plan period. New institutions, such as retailers' buying co-operatives, will supplement K.N.T.C.'s role in Kenyanization. The Government will also ensure that locally manufactured goods will, as far as possible, be distributed by Kenyan traders" (Development Plan 1974 – 1978, Part I, p. 366).

As the diversification of the economy continues, so the population groups in areas which are not engaged in agriculture, above all the urban population, grows. Thanks to the need for provisions for these groups of the population trade continues to increase, and at the present time retail and wholesale trade already amounts to about 10% of the domestic product. Since Kenya follows a liberal economic policy based on the market economy there are no bottlenecks in supply. Trade is well organized in the central places dispersed throughout the country, and the range of goods on display extends from simple items for daily use available in the rural dukas of the trading centres to luxuries for the most refined tastes in Nairobi's specialist shops. But the purchasing power of the African population is very low. Among the poorest sections of the population, it is not sufficient even for the desirable range of vitally necessary goods.

In the rural areas some dukas also stock a few popular medicaments, whereas in the larger towns a wide range of medicines is offered in well-run dispensaries, which do not differ in any respect from those of Europe. It is necessary to import many goods to meet the private needs of the new African middle and upper classes, the European and Asian populations, and of tourists visiting the country. However, it has proved possible to lower the proportion of imports of various commodities considerably by import substitution.

Characterising Kenya as a developing country, Table 9 shows more than half of the annual imports occurring in the fields of industrial supplies, fuel and lubricants, machinery and other capital equipment, transport equipment and consumer goods. The export structure of Kenya is equally typically that of a developing country poor in resources.

According to table 55 of the Statistical Abstract of Kenya 1975, agricultural products together amount to almost 60% of total exports including three principal

Table 9. Total exports and imports 1974 by broad economic category

	Exports	Imports
Food and Beverages	82,101	24,467
Industrial Supplies (Non-Food)	67,807	145,443
Fuels and Lubricants	45,624	81,147
Machinery and other Capital Equipment	2,009	39,642
Transport Equipment	1,188	41,362
Consumer Goods not Elsewhere Specified	12,330	33,178
Goods not Elsewhere Specified	223	1,121
Total	221,282	366,360

Source: Statistical Abstract 1975, Tables 60 and 69

export crops such as coffee (23.6%), tea (11.9%), and sisal fibre (and tows) (10.4%).

A considerable structural change in export production, which is of the greatest importance for the welfare of the farming and herding population, has taken place. Whereas towards the close of the colonial period more than four-fifths of the agricultural exports were produced on large farms and plantations, nowadays smallholders —and to a certain degree even the herdsmen—participate in export production.

V. Transport and Tourism

1. Transport

The rapid development of transport in Kenya is of increasing importance in the transmission of infectious diseases, particularly the diffusion of epidemics. Before an outbreak of some epidemic in a rural area can be tackled by counter-measures, infected persons may well have travelled over great distances and been in contact with many fellow travellers. When smallpox has broken out, for example, the appearance of control points on the roads, run by the police, have been by no means unusual. Unvaccinated persons wishing to continue their journeys have had to submit to vaccination on the spot.

a) Railways: Development of modern transport in Kenya began with the building of the Mombasa-Kisumu Railway between 1896 and 1902. The railway was the pre-requisite for the settlement of the White Highlands by European settlers (see Weigt, 1955, p. 335 and 1932, p. 61). Between the two wars the branch lines to Kitale, Thomson's Falls, and Nanyuki were built, so that the White Highlands were thereby linked with Mombasa and the wider world. This favoured the spread of diseases, especially of plague along the railway line from the beginning of the century until the forties (see Chapter C.I.5a).

Both the railways and the ports of East Africa were under the administration of the East African Railways and Harbours Corporation, until in 1969 the corporation was split up; the railways now belong to the East African Railways Corporation. Over the period 1967–1972, it recorded an increase in passenger traffic in Kenya which was reflected in a rise of income from K£ 1.1 million in 1967 to K£ 1.2 million in 1972. Over the same period the income from freight fell, however, from K£ 15.9 to K£ 15.8 million. This illustrates the competition of road transport, since the total volume of traffic rose during the period from K£ 24.4 to K£ 28.8 million as a result of the country's economic growth (Economic Service 1973, Table 8). In this regard it should be remembered that due to political events and the resulting economic recession in Uganda, transit traffic from Uganda to Mombasa underwent a considerable reduction.

b) Roads: By African standards Kenya enjoys a well-developed road network. In 1972 the primary system consisted of 11,663 km of road, almost half of which were asphalted. Compared with 1965, the extent of roads had been doubled. If the 35,105 km of roads in the secondary system are included, Kenya has a total of more than 46,763 km. A comparison of the road mileage with the number of inhabitants or size of the country is less impressive evidence, since the network density depends on the population density. Whereas it is very closely knit in the densely settled areas of the Central Province, the road network in the sparsely settled North and North East consists of only a few main roads.

Bus services are of the greatest importance in the transport of the African population. Some large bus companies maintain regular services on routes connecting practically all places of importance in Kenya. Thus the mobility of the African population has increased considerably. Full buses, the travellers' luggage piled high on the roof and the whole shrouded in great clouds of red dust, are to be found cruising along the murram roads in even the remoter districts.

Over the period 1963 to 1971 the number of lorries, trucks, heavy vans and motor cars has also increased by around 50 per cent.

The development of road transport is being forcefully promoted and from 1968 to 1973 "some 30 per cent of the total Central Government Development Budget was spent on new road developments" (Development Plan 1974–1978, Part I, p. 345).

In spite of all these efforts it has not yet proved possible to connect all Kenya's settlement to the network of all-weather roads.

During the rainy season a great deal of damage is done to the roads, and many sections become impassable for normal vehicles. The transportation of patients over longer distances by ambulance is quantitatively of no importance by comparison with that of public transport and taxis. Supplying medicine to hospitals and other links with health facilities probably encounters difficulties less as a result of bad roads than of the unavailable or unserviceable utility vehicles or the general shortage of facilities. Acute situations may still be met by the "Flying Doctor Service".

Thanks to police enquiries in every single case, traffic accidents are undoubtedly the health risk best covered by documentary evidence. During the period 1961–1972 the number of traffic accidents rose from 3,573 to 6,613, i.e., by 85%. The number of casualties increased by 161% from 4,030 to 10,518 while, during the same period, the population increased by 30% and the number of motor vehicles by 50%.

Spatial distribution of course shows concentration in towns and along main roads, which implies that population living away from such roads is considerably less exposed to risk. Pedestrians and passengers make up one-third of all deaths resulting from traffic accidents and thus pay most often with their lives (Raval, 1974).

Schram (1968) estimates that there are 55 traffic casualties per million vehicle miles in Kenya, as compared with 65 in Uganda, 6 in the U.S.A. and 13 in the U.K. The behaviour of drivers and pedestrians as well as the condition of vehicles and roads are all probably equally to blame for these high figures.

c) Waterways: In Kenya inland waterways are of no importance, since the rivers are not navigable. The only inland port worth mentioning for its economic significance is Kisumu on Lake Victoria. It serves the passenger and goods transport between the adjoining states. Maritime transport, that is export and import trade and passenger transport, is mainly organized by foreign shipping companies which have their offices in Mombasa. Throughout the decade 1962 – 1971, the number of passengers showed a considerable decline, namely from 79,000 to 23,000 persons. Commodity exports, however, have more than doubled over the same period, showing a rise from 1 million to 2.4 million tons; imports have shown a similar pattern, rising from 1.8 million to 4.0 million tons during the same period.

Sailing ships have lost almost all their economic significance. Whereas in 1962 as many as 858 sailing ships, chiefly dhows, called at Mombasa, Lamu, and other small ports on the Kenyan coast, in 1971 only 235 landings were registered.

Considerable epidemiologic importance attaches to this movement of sailing vessels, which operates chiefly between Kenya and the North African and Arabian countries, insofar as there is a continuous threat of crews importing diseases, above all smallpox, cholera and plague. In earlier years this was a great problem, whereas nowadays strict controls are carried out in the ports of Mombasa. Even so the danger has by no means been eliminated.

d) Air transport: The rapid development of air transport is of special significance for Kenya. It has helped to make tourism one of the most important branches of the country's economy. For a relatively low price and in the short space of 8 – 10 hours hundreds of thousands of tourists arrive from Europe by air each year. Starting from Wilson Airport near Nairobi, charter planes fan out to numerous air-strips in all parts of the country. Tourists can visit remote areas without any difficulty.

The task of the police and administration to maintain law and order in the vast but sparsely populated arid regions is now significantly easier. Valuable help is provided by the flying doctor service of the African Medical and Research Foundation, the aircraft of which are able, on radio request, to send medical personnel and medicines to the remotest parts of the country in a very short time.

Since 1946 Tanzania, Uganda, and Kenya have run their own airline, East African Airways. During the period 1962 – 1971 the transportation of passengers by East African Airways more than doubled to over 1 million

passengers, while in 1971 the load carried trebled to reach 106 million tons/km.

Embakasi, Nairobi's airport, is the most important airport in East Africa, servicing over one million passengers a year. Mombasa, Kenya's second international airport, is presently being enlarged.

Currently it services about 200,000 passengers annually, but the rapid development of tourism on the Kenyan coast will increase the importance of international flights and air tourism to Mombasa.

Regular air services connect all the important towns of East Africa. Internal flights in Kenya not only use Nairobi and Mombasa but also the airports at Kisumu and Malindi. According to the Development Programme numerous air strips in all parts of the country are to be provided with improved and asphalted runways.

2. Tourism

Scarcely any other country in Africa possesses better and more comprehensive pre-conditions for the development of tourism than Kenya. Coastal beaches with their fine sands protected by coral reefs offer ideal opportunities for recreation to tourists from the affluent industrial countries of the northern hemisphere during their northern winter; the game parks excel in their scenic beauty as well as in the great variety of wildlife, so fascinating to the visitor. The infrastructure is of a remarkably high standard, so that the tourist is offered both superior hospitality in modern hotels and various means of transport to reach the parks, which themselves are sufficiently large to allow the tourist to devote himself to the undisturbed experience of what appears to him to be unspoiled African nature. Private initiatives and state economic policy have exploited these locational advantages with skill. The number of overnight accomodation in the hotels rose from about one million in 1965 to 2.5 million in 1972, and tourism is already classed with the main sources of foreign currency earnings. According to the 1974 – 1978 Development Plan, Part I, Table 18.2, tourism is to be even further developed as shown in Table V (Annex).

As a rule health risks are few for the tourist. The most important preventative measure is that against malaria, even though the control of mosquitoes is fairly intensive in the tourist zones of the coastal region. Under appropriate exposure it is considerably easier to acquire venereal diseases. By comparison with many other countries, even those of southern Europe, the standards of hygiene in the tourist hotels is excellent. As a result experience of diarrhoea of various types and causes is rare in Kenya. However, hepatitis constitutes a growing problem even in the field of tourism.

The tide of tourists from the affluent industrial nations to a country in which the majority still lives at the very minimum level, poses problems in many respects. The habits of tourists on holiday sometimes offend the indigenous population on the coast. Prostitution and the occurrence of venereal diseases have risen to the same degree as crime.

The root of the problems remains this. Are the investments made in the infrastructure with the aim of promoting tourism profitable enough for the country at large? Is it justifiable in the face of an increasing land shortage to keep such large areas out of settlement use?

It is the case in Kenya that the national parks are almost without exception located in areas unsuited to agriculture and that the natural potential of the country's pastoral areas is not yet fully utilised. As in other branches of the economy, government policy also aims at careful Kenyanization of the tourist industry, which is at present still dominated by foreign enterprise.

B. Health Services in Kenya

Introduction

A regional medical monograph which seeks to present the reciprocal interactions between man and his physical, biological, and social environments in the modern sense of the word as defined by H. J. Jusatz (1963), i.e., as a medical ecology in the broadest sense of the word (Diesfeld, 1969), has to allow ample space for the health services of a country and their analysis.

The disease pattern and the state of health of a country's population are just as much the result of the health services as they are of the climatic conditions, food production, and distribution of other economic and social variables which make up the general conditions of life. The states of health, economic resources, and the health service, together with the physical, biological and social environments, are bound up with each other in a feed-back system.

1. The Development of Health Services in Kenya until Independence in 1963

Kenya first came into contact with Western medicine in connection with the spread of missions and under the influence of the British colonial administration. Whereas in Uganda the famous Mengo Hospital had been founded as early as 1897 by Dr. Albert Cook, Kenya experienced the beginnings of medical missionary activity only in 1907 when the Church Mission Society (CMS) began its work in Kikuyu. In spite of certain anti-missionary feelings within the British colonial administration, this activity was considered a valuable basis for all medical activities. The African population accepted the missions first and foremost because of this medical activity.

The medical service of the colonial administration concentrated chiefly on the control of the then prevailing major infectious diseases like plague, malaria, and sleeping sickness, which posed the greatest threats to colonization, and on curative services for the European settlers. In 1913 the first Public Health Ordinance was passed. During World War I the African auxiliary troops and the African Carrier Corps were subject to an alarmingly high mortality rate caused by malaria, typhoid fever, and

"exhaustion". Medical examinations carried out among the recruits then provided insight for the first time into the state of health of the African population, and it must be recorded that 50% of the recruits had to be turned down on medical grounds. With the establishment of the "East African Native Medical Corps" at that time, the foundation for future African medical personnel was laid. African youth provided evidence against those attitudes which called into question their capacity for education. Against considerable odds Major Keane attempted to make use of the opportunity. His recommendations had already been followed up at Makerere Training College, Uganda, in 1925, at a time when Kenya was making no efforts in that direction apart from the training of dressers at the Kikuyu Hospital of the Church Mission Society (CMS). However, the growing influence of the new African political leaders provided an impetus for the debate over education and training in general, and for medical training in particular (Beck, 1966).

The Public Health Ordinance of 1921 laid increased stress on the control of malaria and sleeping sickness.

Although the mission hospitals improved their standards and the chief epidemic diseases were brought increasingly under control, the establishment of a health service was delayed again and again. Only late in 1945 was a Ten Year Programme for the development of health and hospital services recommended with a financially generous endowment; it also received substantial support as a consequence of further political developments in Kenya.

With the exception of the Royal Sleeping Sickness Commission set up in 1906, medical research in East Africa did not commence until the end of World War I. In the forties, definite results were obtained in the field of tropical medicine at the Medical Research Laboratory in Nairobi. These efforts were extended by the East African Bureau of Research in Medicine, subsequently taken over by the East African Medical Research Council in 1952. In 1961 this organisation came under the control of the East African Common Services Organization (E.A.C.S.O.), now the East African Community.

Throughout the colonial period medicine and health services in Kenya very much depended on the prevailing political and economic situation in the mother country and its actual interpretation by the colonial administration *.

The revised Public Health Ordinance of 1962, determining the outlines of the organisation of the health services, was arrived at during the preparatory phase leading up to independence and within the framework of the 5-Year Plans since 1963, was only gradually adapted to changing needs (Public Health Act, 1963, Revisions 1964, 1970. Laws of Kenya, Government Printer, Nairobi).

Without doubt traditional African medicine still plays a considerable role among sections of the population. Even allowing for the fact that western medicine has won for itself a considerable reputation over more than three generations, it is nonetheless quite evident that large areas of individual health problems still remain largely in traditional hands.

The impressive figures of calls on ambulatory and stationary medical establishments should not mislead the reader; particularly in the fields of midwifery, pre- and postnatal care, as well as in those of nutrition and health counselling, contact with and the influence of modern preventive medicine are still only marginal over large parts of the country. This state of affairs is expressed as a natural declining gradient in direction to the more remote groups of population; this is especially the case with the nomadic herding tribes of Kenya, which are to be considered as marginal groups in every respect.

Among the groups of population enjoying good access to health services and subject to a rapid process of acculturation, demands for a modern comprehensive health service and for individual medical care are continuously on the increase. But even at the present time modern and traditional medicine continue to find themselves in a certain competitive situation. Although there is still a lack of thorough investigations into the alternating effects of both of these systems of medicine, the practicing doctor will not fail to be aware of the influence of traditional medicine on both his patients and his work.

Van Luijk (1974) has provided a good introduction to this problem of the reciprocal effect for the medical student, pointing out how important it is to take account of the existence of traditional medicine and to come to terms with it as a continuingly integral part of a traditional society in the process of change in a way that proves beneficial to all concerned.

The dichotomy of customs, usages, and magic of a tribal society in relation to health, and the matter of fact necessities of scientific medicine and hygiene, as laid down in the national laws concerning health, are closely connected with the problems of domestic and educational policies of what is not only a multiracial but also a multi-tribal society. These problems not only existed during the colonial period, as has been demonstrated by A.

Beck (1970, op. cit.), but have continued to exist since independence.

Without renouncing tribal ties and traditions, no multi-tribal society can be forged into a nation stable not only in itself but also towards the world outside; without tolerance towards these forces, however, man and his group lose their identity. "Traditional" and "modern" medicine acquire a highly political accentuation in this context.

2. The Structure and Organisation of the Health Services since 1963

In 1963 the Republic of Kenya attained independence. A public health policy was formulated on the basis of the constitutional obligation of the state (The Constitution of Kenya, Kenya Gazette Suppl. No. 27, Acts No. 3, 18th April, 1969) and in the sense of Kenya's "African Socialism" (African Socialism and its Application to Planning Kenya, Government Printer, Nairobi, 1969) within the framework of the Five Year Development Plans of 1966 – 1970, 1970 – 1974, and 1974 – 1978 (Republic of Kenya, Development Plan 1966 – 1970, Nairobi 1965; ibid. 1970 – 1974, Nairobi 1969; ibid. 1974 – 1978, Nairobi 1974).

In the first development plan the policy was defined as follows: ". . . Human health has a major role to play in economic development. That there is a direct relationship between health of a population and its productivity is self-evident and has been demonstrated in the industrial countries, which are now benefiting from the years of investment in health. Apart from the economic benefits, it is incumbent on any government devoted to social welfare of its citizens to provide adequate health facilities . . ." (cited in Onyango, 1974).

In the second development plan 1970 – 1974 (pp. 489 – 490) the tasks are presented and specified in a similar global manner:

. . . "Health Policy and Strategy

The objectives and priorities for health planning must take account of some fundamental factors: an average population growth rate exceeding 3 per cent per year, rapid expansion of urban centres at about 6 per cent a year, wide disparities in the distribution of health services, a severe shortage of medical manpower, varying degrees of efficiency in administration, and financial limitations. There are also difficult issues such as the relative emphasis to be placed on preventive as against curative services, the most effective use of skilled manpower, and the co-ordination of public and private health services. Health planning has thus far been handicapped by inadequate statistical information on the incidence of diseases and the impact of various health programmes, but this difficulty is being overcome through the work of the newly-established Epidemiological Section in the Ministry of Health, which is organizing the regular reporting and analysis of data from the hospitals and health centres. Armed with increasingly reliable information, health planning will continue on an intensive basis, and it is expected that a health planning unit will soon be established within the Ministry of Health. As a further assistance to planning how best to improve health services, the Government is presently considering the appointment of a Special Commission to examine present objectives, functions, structure, staffing and financing of the nation's health service and to make recommendations for their improvement.

---

* Ann Beck: A History of the British Medical Administration of East Africa, 1900 – 1950, Cambridge, Massachusetts, Harvard University Press, 1970. Most data concerning the development of the health services in Kenya are taken from this very detailed and important work.

Meanwhile, the Plan period will see no dramatic changes in the scope and execution of health services but rather a general improvement in the standard of services through more effective co-ordination and consolidation of existing units and a steady increase in facilities, especially in rural and pastoral areas of the country.

Basic policy elements for the Plan period are:—

(I) Construction of urgently-needed new facilities, to the extent they can be staffed.
(II) A substantial programme of renovating and up-grading existing facilities.
(III) Major investments in training at all levels of medical skills.
(IV) More emphasis on preventive and promotive programmes.
(V) The Central Government to take over county council health services.
(VI) Substantially increased assistance to church health services . . ."

In accordance with the precepts of African Socialism as set forth in Sessional Paper No. 10 ". . . it is the Government's long-term objective to provide an adequate level of free basic social services to all its citizens . . ." (cited in Onyango, 1974).

The political decision of greatest importance in this respect was the introduction in June, 1965, of free treatment for out-patients, free board and lodging for all, and free hospital treatment for children under 18 years of age. This led to considerable overloading of the medical facilities without at the same time any corresponding increase in means for staff, buildings, equipment, and medicaments. In spite of all the good-will and readiness that can be observed time and time again among the grossly overworked staff, the resulting disproportionate workload does lead inevitably to a lowering of the quality of their performance. Until late in 1969 the running of the basic health services was the responsibility of the local authorities (County Councils).

Over the course of years "Harambee Self-help activities" have led to the construction of numerous dispensaries and health centres at parish level, which could, however, neither be adequately staffed nor financially and materially supported, thereby providing a burden and disappointment to the population, once the enthusiasm of the pioneer phase had waned. It soon became clear that the provision of such basic health services was too much for the local authorities, and that a central strategy, coordination, standardisation, and planning of services were necessary. A second political decision was made as a sort of urgent measure:

From January 1st, 1970, responsibility—and with it all the financial, material, and staffing burdens—was removed from the County Councils and taken over by the Central Government, i.e., the Ministry of Health. Only the five large municipal councils retained their relative autonomy in the matter of health services.

Both of these decisions had a highly significant influence on the health services. The resulting implications have been discussed by Vogel (1970) and shown to be an enormous additional burden, particularly for the administration of the headquarters, but also a unique opportunity for re-organisation and standardisation of administration and medical measures.

In addition the recent years since independence have seen the re-arrangement of the organisational structure in some fields of the health services. The present state can be described as follows:

At the political level, the Minister is responsible to the President and Cabinet; at the administrative level, the permanent secretary is the highest administrative official. In the Headquarters itself, a professional medical branch is under the Director of Medical Services, and an administrative branch, under the Deputy Secretary.

The Director of Medical Services (DMS) and his deputy DMS are supported by six assistant directors of medical services—for Personnel, Hospitals, Preventive Services, Laboratory Services, Recruitment and Training—as well as a Chief Nursing Officer.

On the periphery, the organisation follows the general administrative hierarchy. The Provincial Medical Officer carries full responsibility and authority for all medical and related administrative and financial matters within his province. At the lowest district level the District Medical Officer is answerable to the Ministry of Health by way of the Provincial Medical Officer (see organogram).

But still "three major constraints are at present hampering an accelerated achievement of the health goals, an unsatisfactory level of service in rural areas due to insufficient service delivery points, inadequate resources and organisation and a shortage of manpower, particularly of paramedical staff. The 1974 – 1978 Plan is designed to remove these contraints as rapidly as possible within the overall contraint of national resources available for health" (Development Plan 1974 – 1978, Part I, p. 449).

In comparison to previous strategies . . . "two major corrective measures have been introduced for the 1974 to 1978 Plan; basic paramedical training will be substantially expanded to the maximum the system can now carry and absorb and secondly the Government will start implementing an integrated and comprehensive master plan for the development of basic rural health services" (ibid. p. 450).

The Ministry of Health is responsible for the following:

* the national health policy
* the national health development plans
* organisation and administration of central health services
* training of health and allied personnel
* health acts and regulations
* promotion of medical sciences and maintenance of medical and health standards
* liaison and coordination with other government departments and nongovernment agencies
* international health regulations

(quoted in Onyango, 1974 op. cit.)

The Assistant Director of Medical Services, Preventive Services is responsible for:

* Division of Maternal and Child Health
* Division of Health Education
* Division of Environmental Sanitation

Organogram Ministry of Health, Kenya (1971)

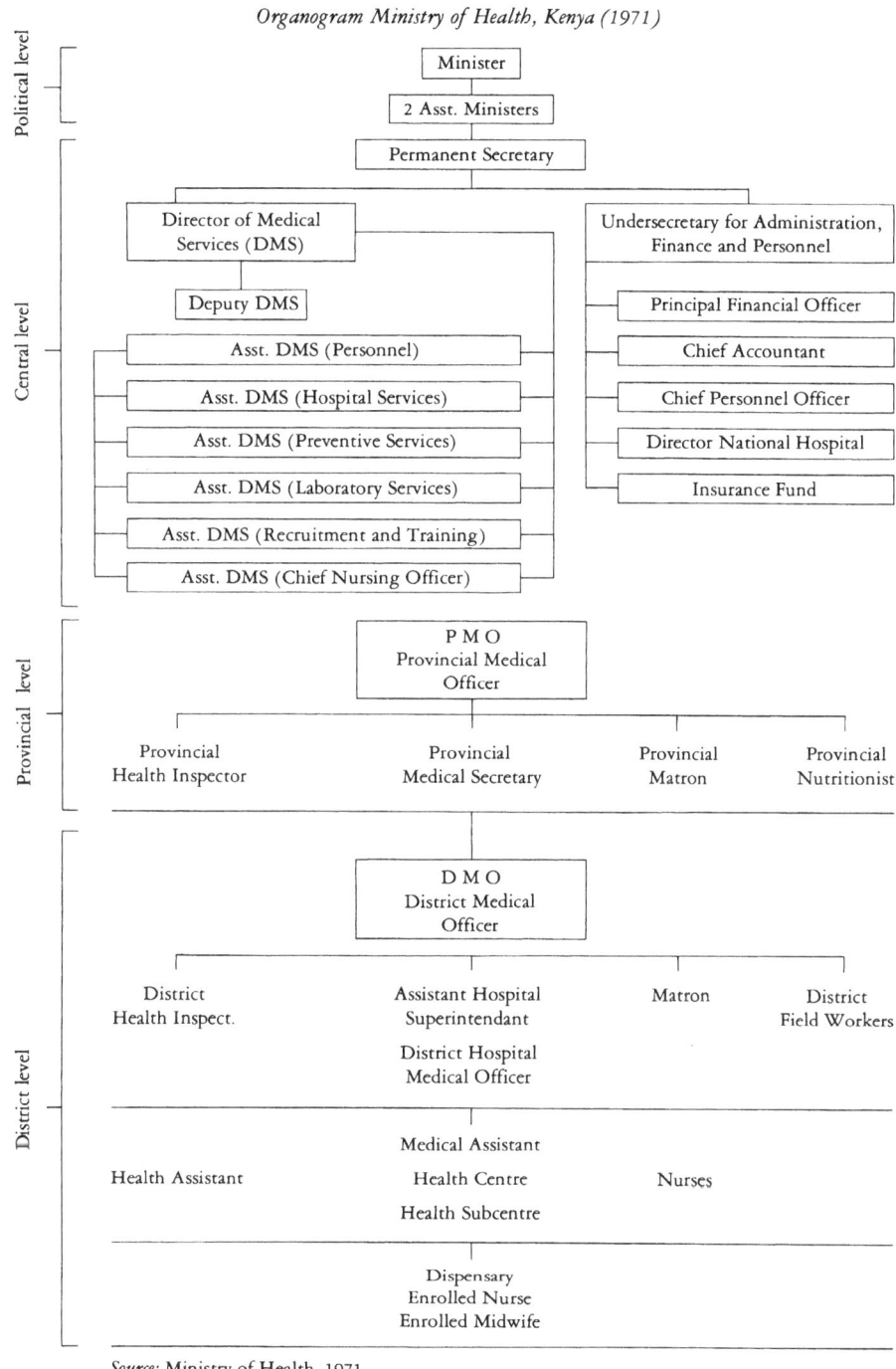

Source: Ministry of Health, 1971

At present he is responsible, moreover, for the functioning of the basic health services until such time as a special assistant director of medical services for this office is appointed.

Public Health Act: Apart from the development plans for the health services, the main legal foundation consists of the "Public Health Act". Based on the first Public Health Ordinance of the year 1921, revised repeatedly and last in 1962 in the course of preparation for independence, the Public Health Act evolved as the national legislation for health; in 15 parts it regulates by law all activities relating to health for which the state is responsible *

* Vogel, L.-C. Health Services and National Development in Kenya, mimeographed Lecture Notes, Medical Research Centre, Nairobi, Department of the Royal Tropical Institute, Amsterdam, The Netherlands, Nairobi 1971. In this work for medical students Vogel describes the health services of Kenya with exemplary detail and clarity. This manuscript served the author as a corrective for quite a number of problems and data which had been acquired in the course of his own investigations and study of original sources.

Besides this central law there exists a number of further laws, which although revised in parts, in the main follow a pattern formulated before independence. Under the particular laws, regulations are issued, approved by the Attorney General, and gazetted. The most important of these are:

The Births and Deaths Registration Act of 1928, revised in 1962
The Malaria Prevention Act of 1929, revised in 1962
The Dangerous Drug Act of 1933, revised in 1962
The Pharmacy and Poisons Act of 1957, revised in 1962
The Food, Drug and Chemical Substances Act of 1965
The Mental Treatment Act of 1949, revised in 1962
The National Hospital Insurance Act of 1966
The National Social Security Fund Act of 1965
The Workmans' Compensation Act of 1948, revised in 1962
The Employment of Women, Young Persons and Childrens Act of 1948, revised in 1962
The Kenya Red Cross Act of 1965
The Kenya Society for the Blind Act of 1956
The Royal Tropical Institute Act of 1965
The Medical Practitioners' and Dentists' Act of 1910, revised in 1962
The Nurses', Midwives' and Health Visitors Act of 1965
The Radiation Act of 1962
The Human Tissue Act of 1956
The Human Anatomy Act of 1967.

Under a variety of laws a number of statutory boards were established, including amongst other the following:

The Central Board of Health, under the Public Health Act
The Medical Practitioners' and Dentists' Board, under the Public Health Act
The Nurses', Midwives', and Health Visitors' Council, under a separate Act
The Public Health (Standards) Board, under the Food, Drug, and Chemical Substances Act
The National Hospital Insurance Advisory Council, under the National Hospital Insurance Act.

3. National Health Budget

Total expenditure in the field of health services during the period 1957 to independence in 1963 remained largely constant at £ 2.5 million annually, i.e., 6.25 sh. a head, assuming an average population of 8 million during these years.

The period 1963/64 was the first fiscal year following independence. Decoding of the expenditure (see Table VI, Annex) shows that more than half of the expenditure went on salaries and wages and individual staff expenses, a tendency which has persisted through the years in spite of salaries which are in themselves low. Expenditure for "medical stores", a term that covers chiefly medicines and expendable items, amount, however, to only 16.7%; the running costs of the ministry add up to 7.6%, and those of medical institutions (i.e., essentially the hospitals) to 14.4% without those under the heading "medical stores". Health centres and dispensaries were still under the authority of local government at this time. A contribution of £ 57,000, i.e., a mere 2.4% of the Ministry of Health's total expenditure, went towards the cost of running the mission hospitals.

Returns for services during this period, however, amounted to an average of £ 280,000 at a time when treatment was in no way free for all, and thereby reimbursed about 12% of the total expenditure (Ministry of Health Annual Reports 1964 – 66).

The year 1965 saw the political decision on free treatment for all out-patients and free hospital treatment for children. This law took effect on June 1st, that is at the beginning of the fiscal year 1965/66.

This caused a considerable financial burden, directly abolishing income amounting to K£ 280,000 and adding an extraordinary allocation of K£ 55,000 to support the mission hospitals (since these had been affected by the law on free treatment for out-patients), and indirectly by causing a rapid rise in the number of treatments by 50 – 100% (see below), which had not been provided for in terms of staff and materials.

Only with the expansion of the budget of 1968/69 did the volume of the health services budget also rise; compared to that of 1967/68, the increase amounted to more than K£ 300,000, i.e., 8.75%, which was twice the rate of the budget in general (+4.3%). From 1969/70 the health services budget rose in proportion to the main state budget (see Table VI, Annex).

The major part of the increase in cost was due to general price increases, the creation of new jobs in the course of development projects, and in the course of salary adjustments following promotion to higher grades. The individual headings of the health services budget remained substantially the same, as did their shares in the overall budget (see Table VII, Annex).

The Kenyatta National Hospital, now set apart as a national referral and teaching hospital of the Medical School at the University of Nairobi, is entered in the budget under its own heading. This shows that in 1973/74 this hospital claimed K£ 678,000, that is 7% of the total expenditure, not counting the staff costs embodied in appointments of ministerial personnel; set against this the remaining 50 or so government hospitals, again excluding staff costs, received 6.2% for operating costs and 11.5% for materials and medicaments, i.e., a total of K£ 1.6 million. A small sum is set aside for the Mother and Child Health Services.

In 1968/69 the health budget was increased by approximately K£ 2 million, chiefly on account of increased wages and employment and the expansion of the Kenyatta National Hospital.

From January 1st, 1970, the institutions of the peripheral health services, the health centres and dispensaries, which had hitherto been under the administration of the local authorities and had been extended during several years of "Harambee Self-help" activities, were taken over by the central government and entered in the health ministry's budget under their own budgetary heading (No. 103; Estimates on Recurrent Expenditures 1971/72, Nairobi 1971, p. 49), accounting for about 20% of the overall budget. During the initial years the budget was made up of the current expenses for operation and medicaments, together with the staff salaries, which amounted to K£ 752,000, i.e., 58% of this item in 1970/71. Expenses for materials and medicaments amounted to

K£ 361,150, i.e., 28%, and operating costs to K£ 182,850, that is 14.1%.

From 1973/74 this entry was reduced by the staff costs of the health budget, which were transferred to the heading of overall staff expenditure.

Assuming that in 1973/74 the amount set aside for staff expenditure in the peripheral health services remains the same as in previous years, it will again amount to about 12% of the total expenditure of the Ministry of Health.

The estimates of 1973/74 in the health ministry's composite budget were cut by a net sum of K£ 87,000, that is by 10%, which corresponds with a gross increase of less than 0.1% compared to the previous year, whereas the budget as a whole experienced an expansion of 10% compared to the preceding year (see Table VII, Annex).

This is the first reaction to the generally unfavourable economic development, marked by an increasing deficit in foreign trade.

With regard to the population increase and the various legal re-assignments of competence, the per capita expenditures of the Ministry of Health increased from 7.14 sh. in 1967/68 to 15.19 sh. in 1973/74, with a retardation in the rate of increase after its peak in 1970/71. In 1973/74 the per capita expenditure decreased for the first time, in accordance with the cut-back in the budget.

In these figures no allowance has been made for the expenditure by local government bodies, by the churches and other supporters of health facilities of the "Health Development Expenditures", all of which would have to be taken into account should a correct picture of direct expenditure on health and care for the sick be attempted.

Although supported by grants from the central administration, district hospitals remained under local government authority until 1964, and health centres and dispensaries, until 1970. As a rule local authorities were overburdened in both these fields. A take-over by the central administration provided financial relief, but there was not necessarily any improvement in the performance of the facilities, because in the early stages the central administration was likewise unable to cope fully with the organisational, personnel, and material expense.

The Development Plan 1970 – 1974 repeated the demands concerning the responsibility of the government for an accelerated development of the health services and for the availability of means for this within the framework of the development plans.

At the same time there was a request that about 40% non-state controlled activities concerning health, run by churches and other private organisations, should be subject to increased control and integration. To this end a commission had been set up as early as 1966 with the task of making appropriate recommendations. The basic policy 1969/70 provided the following elements (Development Plan 1966 – 1970, p. 490):

1. Construction of urgently needed new facilities to the extent they can be staffed
2. A substantial programme of renovating and up-grading existing facilities
3. Major investments in training at all levels of medical skills
4. More emphasis on preventive and promotive programmes
5. The Central Government to take over county council health services
6. Substantially increased assistance to church health services

The municipal and country councils too were included in the development programmes of 1965/69. Apart from including directly related health and social projects they also applied themselves to indirectly related ones, such as water supplies, sewage disposal, housing programmes etc.

However, the "Economic Survey 1973" noted that the positions provided for in the plans could not be fulfilled, although the means had been made available (Republic of Kenya, Economic Survey 1973, Central Bureau of Statistics, Ministry of Finance and Planning, Nairobi 1973, pp. 54–57).

The Health Development Plan 1974 – 1978 for the first time mentions rural health services development in the first place and hospital development in the fourth place, although the latter still receives the largest share (see Table 10).

4. Church Health Services and other Voluntary Societies

a) Church health services: The missions were the first bodies in Kenya to attempt the medical care of the population, and that as a means of evangelisation. While the colonial administration was looking after the care of the settlers and their labour, as well as undertaking the fight against epidemics, the missions started their activities in the "native reserves". This division remains recognisable in the pattern of distribution of the mission hospitals.

Relations between the missions and the colonial administration were by no means untroubled, although the competences appeared to be clearly divided. As the colonial administration began to feel responsible for the medical care of the population, increasingly so after World War I, conflicts developed, especially where standards and the quality of the missions' medical work were concerned; these involved certain administrative and professional controls which the missions were not prepared to accept. A point of particular dispute in the mid-twenties was female circumcision among the Kikuyu, as noted in the study by A. Beck (1970, op. cit.). Although the colonial government greatly promoted the development of its own medical care during the twenties, it was not able to dispense with the services of the missions in this field. It therefore granted official, though small, financial support; during the years of the recession even this was considerably reduced. After World War II a better understanding between both sides was achieved; the missions accepted certain limitations concerning their competence, together with the requirements and controls of the health authority. On its part the authority granted financial support to the medical work of the missions, albeit without keeping pace in the least with the

Table 10. Health Development Programme 1974 – 1978: Programme Summary

(K£'000)

Programme	1973/74	1974/75	1975/76	1976/77	1977/78	Plan Total
(a) Rural Health Services—						
Expanded Programme	—	370	530	1,290	1,825	4,015
Family Planning Programme	—	735	1,020	717	—	2,472
	—	1,105	1,550	2,007	1,825	6,487
(b) Health Training—						
Regular Programme	416	503	678	670	605	2,872
Family Planning Programme	14	830	726	273	130	1,973
	430	1,333	1,404	943	735	4,845
(c) Public Health—						
Environmental Sanitation	63	100	153	183	243	742
CDC/VBDC *	30	41	60	60	60	251
Health Education	—	7	75	90	70	242
Family Planning	—	188	94	195	59	536
Other	—	14	7	7	7	35
	93	350	389	535	439	1,806
(d) Hospital Development—						
New District Hospitals	503	612	565	770	835	3,285
Extensions, Improvements	409	427	402	722	944	2,904
Staff Housing	122	271	296	358	479	1,526
Provincial Hospitals	21	200	300	637	1,383	2,541
Kenyatta National Hospital	600	604	1,210	1,210	900	4,524
	1,655	2,114	2,773	3,697	4,541	14,780
(e) Medical Stores	—	20	70	70	50	210
(f) Grants-in-Aid (Municipalities)	—	25	25	25	25	100
(g) Research	—	33	95	95	60	283
Total	2,178	4,980	6,306	7,372	7,675	28,511

* Communicable Disease Control/Vector-borne Disease Control. Source: Development Plan 1974–1978, p. 453, Nairobi 1974.

increase in services. In 1959, a joint commission made the following recommendations (quoted in Onyango, 1974 op. cit., p. 123):

To recognise the important services rendered by the medical missions and give support for continuation and expansion,
to set up the Central Advisory Board at the national level,
to organise a mechanism of co-ordination at district level,
to aim at providing two beds (government and mission together) per 1,000 population,
to support medical missions financially.

These recommendations were by no means fully implemented, grants remained small, the threat of financial collapse of the medical mission activity increased continuously—particularly since the performance of the state health services after 1966 forced the missions to follow suit. During the period 1961/62 – 1966/67 the "recurrent grant in aid" remained at a stand-still with K£ 55,883 annually, whereas a minimum of K£ 175,000 would have been required by 1966/67 (Ministry of Health Annual Report 1966, Nairobi, 1971, p. 139).

The Development Plan 1966 – 1970 noted that without greater financial aid by the government the collapse of medical mission services had to be expected; by this stage they constituted 46% of all hospitals, and 23% of all hospital beds, and were thus an indispensable factor in the medical care of the population. In addition 36% of all enrolled nurses and 40% of all enrolled midwives were trained in mission hospitals. In 1967 mission hospitals employed 60 physicians. The latest figures for the year 1966 run to totals of 700,000 out-patients and 90,000 in-patients handled by all the mission hospitals together.

In the remote areas of the north, mission hospitals were also charged with public health responsibilities, as for instance in the Turkana District or in the southern part of Marsabit District, where the health authorities confined themselves to occasional control visits (Ministry of Health Annual Report 1966, Nairobi, 1971, p. 95).

The Central Advisory Board for the study of possibilities of a "closer integration of the health services of Kenya" (Ministry of Health. The closer integration of

Additional material from *Kenya*
ISBN 978-3-642-66937-8 (978-3-642-66937-8_OSFO1),
is available at http://extras.springer.com

the health services in Kenya. Report of a sub-committee appointed by the Central Advisory Board on Medical Missions, Government Printer, Nairobi 1968), first appointed in 1966, published its report only in 1968 and stressed the necessity of closer cooperation and coordination. Overlapping and competition between the services were criticized. Since the ultimate aim, the full take-over of all medical services, was not possible at that time on account of the financial burden connected with it, the following recommendations were made:

". . . to lay down staffing establishments
to promote free interchange and secondment of staff
to establish comparable terms of service for all staff
to lay down national standards of staffing, equipment and accommodation for nurses and midwives training schools
to carry out a nation-wide survey to decide on needs for facilities and services per district
missionary medical workers should join in preventive medical work
to standardise hospital practice and procedure
that current grants be given by Government to enable church hospitals to charge the same fees as government hospitals and to maintain a comparable standard of service
that grants for recurrent expenditures be based on the criteria of workload, staffing and training . . ."

Following this the annual grants in aid for the mission medical services were noticeably raised:

In 1968/69 a supplementary budget raised a grant of K£ 58,610 to K£ 93,610, from 1971/72 it rose to K£ 115,000, and from 1972/73 to K£ 152,760 (Republic of Kenya, Estimates on recurrent expenditures of the Government of Kenya 1968/69 to 1973/74, Government Printer, Nairobi).

b) Voluntary societies: Besides a number of smaller self-help organisations run on a communal basis, there exist larger charitable organisations that receive a considerable part of their income from outside the country. Such a case is the *Kenya Society for the Blind,* which is attached to the Royal Commonwealth Society for the Blind and is financially supported by it.

Seven mobile opthalmic clinics visit health centres, dispensaries, and hospitals, and in cooperation with the Flying Doctor Service a large number of eye operations are carried out by ophthalmic specialists in remote hospitals. A part of the nursing staff is supplied by the Ministry of Health. There are five primary schools and one secondary school specifically for the blind. The *Kenya Society for Deaf Children* is supported by a number of voluntary organisations and carries out medical as well as educational work. At the present time there are 18 schools for the deaf.

The *Family Planning Association of Kenya* operates in Kenya with the approval of and in agreement with the Ministry of Health; it receives significant support from the IPPF, runs several clinics, and untertakes advisory and information work. About one third of all women in Kenya who take part in family planning programmes are cared for by this organisation, the remainder by the Kenya National Family Planning Programme.

Probably the largest voluntary society in Kenya is the *African Medical and Research Foundation* (A.M.R.F.), which incorporates the East African Flying Doctor Service; founded in 1957. It is supported by its members and several national branches in various countries, as well as by contribution from the governments of Kenya and Tanzania.

While in the beginning emphasis was largely put on the now famous Flying Doctor Service, over the last few years work has concentrated around the problem of strengthening rural health services and the training of personnel to undertake these services. A.M.R.F. comprises today a rural health department, which organises mobile medical teams for various specialities, especially surgery, ophthalmology, preventive medicine and health education. A health education department mainly engaged in the training of public health coordinators and the production of manuals and other health related papers. A training, publication and printing department engaged in refresher courses for health workers and in the production of teaching material. A certain amount of research is carried out mainly in the field of epidemiology.

Best known is the nursing and radio service, providing a surgical nursing service to approximately 60 rural mission — as well as government hospitals throughout East Africa, accompanying emergency flights, and operating the medical radio network.

Radio communication is an essential service to the rural hospitals, clinics, and dispensaries where they are installed and provides reassurance for the medical personnel.

The aviation service is carried out at present by 6 small aircraft based at Nairobi and 2 based in Tanzania. Apart from single emergency flights, a part of the rural health programmes is airborne; e.g., the government emergency action on the outbreak of cholera in 1972 was successfully assisted by the Flying Doctor Service.

5. Health Service Facilities and their Regional Distribution

a) Hospital services: Unlike the "basic health services" the hospitals are solely concerned with curative services. Corresponding to the functional and organisational arrangement of the health services and the hierarchical organisation of the administrative structure, the government hospitals are also classified as national, provincial, district, sub-district, and special hospitals.

The *Kenyatta National Hospital,* the only national general hospital, provides specialist services on a wide scale. Simultaneously it also serves as a teaching hospital for the Medical School of Nairobi University.

The *Provincial General Hospitals (PGH)* constitute referral hospitals for the seven provinces; they are located in the main town of each province — Nyeri (Central Province), Mombasa (Coast Province), Machakos (Eastern Province), Nakuru (Rift Valley Province), Garissa (North Eastern Province), Kisumu (Nyanza Province), and Kakamega (Western Province). At the same time they act as training centres for paramedical personnel.

Table 11. Hospitals and hospital beds by Operating Agency, 1967

Agency	Hospitals		Beds		Mean number of beds
	number	%	number	%	
Central Government *	74	37	8,076	59	109
Local Government	2	1	153	1	76
Catholic Church Hospitals	58 } 91	29 } 46	2,461 } 4,015	18 } 29	43
Protestant Church Hospitals	33	17	1,554	11	47
Private Hospitals	26	13	1,075	8	41
Company Hospitals	6	3	296	2	49
Total	199	100.0	13,615	100.0	68

* of which eight are prison hospitals, with 338 beds.

Source: Development Plan 1970 – 1974.

At the district level the *District Hospitals* serve as referral hospitals for the basic health services, and the larger ones among them also train paramedical personnel. Such district hospitals exist in all 40 districts, apart from the seven districts in which the provincial general hospitals are situated and in Nyandarua, where the hospital was still under construction at the time of writing.

Over and above this there are a number of smaller general hospitals at the sub-district level, as for example in Londiani, Kapkatet, Molo, Naivasha, Lokitaung, or Makindu.

In addition there are some special hospitals for infectious and chronic diseases such as leprosy, tuberculosis, or mental diseases, together with one military and eight prison hospitals. Because they are only available to special groups it seems right to exclude them from any consideration of regional distribution and provision for the public sector.

The greater part of *non-governmental hospitals* is maintained by the catholic and protestant churches. With no less than 46% of all the hospitals and 29% of all the hospital beds, they make a significant contribution to medical care in some parts of the country (see Table 11 and 12).

A not inconsiderable number of *private hospitals,* often maternity homes or company hospitals found chiefly in the towns, is available to special population groups.

In 1970 the overall bed/population ratio amounted to 1.25 per 1,000, a figure somewhat above the average and not attained by many developing countries. The investigation into the regional distribution, however, shows considerable differences. In the period 1966 to 1970, the number of hospital beds increased from 10,419 to 14,525, by 40%, whereas the population increased by about 30%. The number of beds in governmental hospitals went up by about 15% and those in mission hospitals by about 50%, many of which, nevertheless, could not meet the considerable demand for hospital beds. Of the total number of hospital beds in Kenya, only 13% are pediatric beds, alarmingly few, if one considers the high proportion of children (0 – 14 years of age) making up 48% of the total population (Census 1969).

The statistical data vary according to the sources consulted and even according to individual years, because it can never be unequivocally established which beds and categories have been entered into the individual statistical returns and which have not. Any attempt to evaluate the developmental trend is thus made difficult.

A rough classification of hospitals and hospital beds by province immediately reveals the extreme variability in their distribution, not only in respect to the population as such but also as regards their accessibility compared to the population density and area to be covered (Table VIII, Annex).

Apart from Nairobi, where 11% of all hospitals and over 20% of all hospital beds are concentrated, only the Central, Coast, and Rift Valley Provinces enjoy more than 1 bed per 1,000 inhabitants (Table VIII, Annex).

Bearing in mind that this somewhat unrefined compilation includes very different hospitals of greatly varying size and equipment, and also that the emphasis of medical care is not to be seen in the hospitals but further away on the periphery, it must be noted that a relationship between hospital and population—which may show regional differences of 1 : 50,000 or 1 : 100,000 as the case may be—by comparison with other developing countries, Kenya is characterised by a good average provision.

Table 12. Percentual distribution of beds (excluding cots) by Agency and Province, December, 1970.

Province	Operating agency		
	%	%	%
	Government	Mission	Other non-government
Central	58	41	1
Coast	68	11	21
Eastern	53	47	—
Nairobi	76	1	23
North-Eastern	100	—	—
Nyanza	59	38	3
Rift Valley	51	30	19
Western	35	66	—

Source: From data in Economic Survey, 1971 (totals are for government 8,402, for missions 3,921, for others 1,644).

However, this touches upon neither the question of the accessibility of hospitals to the public nor upon a further one, namely whether, given an improved health services infrastructure and communications, the goal should be even more effective hospital care with fewer, but even larger and better equipped hospitals. This is especially the case when considering that it is extremely uneconomical to maintain hospitals with an average number of just over 100 beds at the district level, or, even worse, in the mission hospitals, with an average of less than 50 (cf. Table VIII).

The attention given to this fact for some years already is expressed in the National Development Programme and among the missions by working more towards the improvement of capacity and quality rather than quantitative increase. Considering the mean area for which a hospital has to cater, it becomes evident that the Eastern and North Eastern Provinces, on account of their large extent and their overall structure, present the least favourable relationship between hospital and area; it should, however, be emphasised that this makes no allowance for population distribution within the region, consideration of which would present the conditions in a more favourable light.

b) Basic health services: The term basic health services embraces those preventive, curative, and promotive services which serve the population beyond the hospitals and do not require such structures and special services as are available only in hospitals. As a standard the WHO recommends what has in the meantime become a generally accepted guideline of one health centre for every 50,000 or so inhabitants, together with one dispensary (a sub-unit of the health centres) for about 5,000 inhabitants.

The health centre is a station for out-patients, equipped with just a few beds for emergency and maternity cases, the staff of which carries out clearly defined tasks of preventive medicine in the course of curative measures (King, 1966). Within the framework of general out-patient consultations, systematic treatment is to be carried out together with the most important tasks of preventive medicine and health education. Separate sessions are held for mothers and their children aged five years or less ("under five clinics"), and at these the mothers are encouraged not only to bring their sick ones for treatment, but also to include their healthy children as well. In this way not only can treatment of the sick children be undertaken, but general vaccinations can be carried out, and observations can be made of the state of health and nutrition of families as a whole, which can in turn be influenced in matters of health education. In addition ante-natal care can be provided on an out-patient basis, followed by obstetrical and nursing treatment during confinement either at the station itself or during home visits. Moreover, the health centre is regarded as a suitable place for advice on family planning, given in an individual and understanding manner.

In this way health centres, and before them the dispensaries, with their staff become the backbone of the health services, particularly in rural areas. These basic health services are coordinated by the district medical officer (DMO), who is also responsible for the coordination of these services with hospitals or such specialised programmes as epidemic control and vaccination campaigns.

On occasion in some remote areas these static basic health services are supplemented by mobile health units, which depend in essence on vehicles capable of moving cross-country.

At present the Flying Doctor Service of the African Medical Research Foundation (A.M.R.F.), a voluntary agency (see Chapter 4 b), provides the only form of mobile unit operating from the air in close and very effective cooperation with the government health services.

Before 1970, the local authorities (county councils) were responsible for the development and management of rural health services. The county councils were thus responsible for the basic health services for 90% of the population. As the demand for services outgrew the resources available, the burden grew too heavy for the councils to carry, to the extent that the Government decided to take over health services from the councils in 1970. The problems which the Ministry of Health now has to deal with are poor and inadequate facilities, understaffing and unsatisfactory standards of services (Republic of Kenya, Proposal for the Improvement of Rural Health Services and the Development of Rural Health Training Centres in Kenya, Republic of Kenya, Ministry of Health, Nairobi 1972, p. 1).

In 1970, 195 health centres and 603 dispensaries were in operation in Kenya, though some of them were inadequately equipped with staff and materials. 88% of the health centres were run by the Ministry of Health, 9% by the municipalities, chiefly those of Nairobi and Mombasa.

Of the 603 dispensaries, more than 62% were maintained by the Ministry of Health, and 2.3%, by the municipalities of Nairobi and Mombasa. In comparison with the health centres, more than 25% of the dispensaries were run by the missions, and nearly 10%, by large firms (see Table IX, Annex).

The data in the various statistical returns and years fluctuate here as well; very much as in the case of hospitals, the definition of health centre, health subcentre, and dispensary are not always clear. Moreover an increasing number of health subcentres were upgraded to health centres, or renamed dispensaries, over recent years, since it had been agreed to employ only these two main categories in the basic health services. In the course of implementing the development plans more and more dispensaries were to become health centres. As in the case of the hospitals, there are great differences in the regional distribution of health centres and dispensaries, and within the provinces concentration on the rural agglomeration areas has aggravated the situation even further, to the detriment of the sparcely settled areas.

Nairobi has a ratio of dispensaries to population of 1 : 15,558, and taken in combination with the 6 health centres, this ratio falls to 1 : 11,020.

In the seven provinces the health centre ratio fluctuates between 1 : 43,000 and 1 : 84,000, with a national average of 1 : 60,000. With an adequately close network of dispensaries the health centre system might be regarded as a referral system, for the dispensaries are well suited to take adequate care of the population, given the prevailing economic circumstances. This would be the case if the dispensary/population ratio were 1 : 5,000 to 1 : 10,000 in accordance with population density. But in fact, it fluctuates between about 1 : 12,000 in the Coast Province and 1 : 100,000 in the Western Province, the national average being almost 1 : 20,000.

In 1972 the Ministry of Health introduced a "Proposal for the Improvement of Rural Health Services and the Development of Rural Training Centres in Kenya" (Min. of Health, Rep. of Kenya, Govt. Print. Nairobi, 1972), in which the idea of a "medical assistant" was reintroduced after years of hesitation; Fendall (1963) had first put forward and successfully introduced it in the early sixties.

The proposal concerns itself with the purpose, functions, organisation, and staffing of six rural health training centres (RHTC) which are to be implemented during the Period 1973 – 1976 and to provide trained manpower for the rural health services through to 1984.

Having analysed the problems, albeit on the basis of insufficient information and projected them to 1984, a list of 34 objectives in the field of family health, communicable diseases, environmental health, nutrition, infections of upper respiratory tract and gastroenteral infections was drawn up. This was contrasted with an analysis of constraints and manpower, together with one on necessary training programmes. On this occasion it was noted that the present coverage for the different age groups of the target population barely attains 20%, and that in the essential areas (i.e., the vulnerable age groups) 50 – 70% were reached, this figure falling to only 33% in some areas.

The population relation of health centres is to be brought up to 1 : 31,000, that is nearer the final target of 1 : 20,000, and that of all basic health services, including dispensaries, up to 1 : 10,000. Regional differences are also be adjusted. Each health centre and sub-centre is to be run by a medical assistant, the dispensaries by an enrolled community nurse. The staff/population ratio is to be improved from the present 1 : 70,450 for medical assistants to 1 : 32,000 (= +220%), the present 1 : 14,765 for community nurses to 1 : 10,348 (= +256%), and the present 1 : 21,908 for health assistants to 1 : 10,348 (= +221%).

The entire project is to be financed nationally, with considerable multi-lateral and bilateral support.

c) Mental health, dental and pharmaceutical services;
Mental health services: Fifteen per cent of all hospital beds in Kenya are located in the two mental hospitals in Nairobi (Mathari with 730 beds) and Gilgil (520 beds).

A few more beds are to be found in the provincial hospitals. The care of the mentally sick is still in its infancy. Knowledge of the sorts of mental diseases and abnormalities, as well as their distribution, rests largely on the experience of these institutions which contain a very select body of patients. Investigations based on the population as such do not exist.

Of the few physicians concerned with the mentally ill only three are trained specialists.

There are very few specially trained nurses. As elsewhere the development of psychopharmacology has reduced the time spent in these institutions and liberated them from the apprehensiveness hitherto associated with them, although conditions in psychiatric care are still far from satisfactory (Mustafa, 1974, p. 453, cited in Vogel, Ministry of Health Annual Report 1966, Nairobi 1971, p. 128).

Dental services: Apart from dental extractions within the framework of government, and possibly other, hospitals, there was no organised dental treatment prior to 1966.

The past ten years have seen the installation of dental stations at least in the provincial hospitals and in the Kenyatta National Hospital (Ministry of Health, Annual Report 1966, Nairobi 1971, p. 132).

The independent dental surgeons, numbering about 70 in 1974, are concentrated in the towns and care for the well-to-do groups of the population. 49 dentists are registered in Nairobi, whereas Mombasa is served by only 9, Kisumu by only 4, Nakuru by 3, and Thika, Eldoret, Kitale, Nyeri and Kakamega by only one each (East African Dental and Medical Directory, 1973, pp. 33 – 40, East African Dental Association, Nairobi).

Although recent investigations contradict the assumption that the teeth of the Africans are in good order—the rapid increase of sugar consumption has already demonstrated its devastating effects—little attention is paid to preventive dental hygiene and dental health care in school, owing to the lack of opportunity (Bakshi, 1974, cited in Vogel, p. 519 et seq.).

Pharmaceutical services: The "Pharmacy and Poisons Act" and the "Dangerous Drugs Act" are both set out under Chapter 254 of "The Food, Drugs, and Chemical Substances Act"; together with the "Public Health Ordinance", laid down in Chapter 242 of the Laws of Kenya, they regulate the traffic in medicines. The latest revision of this law took place in 1970.

Medicaments are classified according to how dangerous they are or on their mode of action. So far there is no pharmaceutical industry. The medicines or their basic contents are prepared and packed from imported bulk constituents at the Central Medical Stores, thereby saving a considerable amount of expenditure.

There is a Pharmacy and Poisons Board under the Ministry of Health, which, by appointment of an Inspector of Drugs, supervises the enforcement of the law. In addition a Medical Stores Advisory Committee advises the Central Medical Stores in matters concerning the make-up of medicaments stocked in Kenya and in particular which medicaments are to be released for use in government hospitals. The Chief Pharmacist, together with his provincial colleagues, is responsible for controlling the stocks of private as well as government hospital pharmacies.

The East African Dental and Medical Directory 1973 (East African Literature Bureau, Nairobi, 2nd ed. 1973), lists a total of 148 pharmacies, 113 of which (i.e., 70%) are registered as being located in Nairobi. Mombasa has 16 pharmacies, the remaining 19 being in places like Nakuru, Kisumu, Eldoret, Kericho, and elsewhere.

The majority of the population is thus dependent on the medicaments available in state or church hospitals, health centres and dispensaries. It is impossible to provide an estimate of the consumption of medicaments; in government hospitals the costs of medicaments amount to about 12% of the overall budget. In its estimated recurrent expenditures of commercial pharmacies the Development Plan 1970 – 74 assumes a figure of K£ 1.3 million, that is 2.3 Ksh. per person. No consideration is given to the free sale of non-scheduled medicaments, free of prescription or "under the counter" sales. The same is true of the complex but quite unknown realm of folk medicine.

Studies by Vogel et al. (1974, 1976), carried out in the out-patients' department of Machakos and Kiambu Hospital, showed that an average of 1.3 sh was spent per patient/visit on medicaments. They studied in detail the problem of medicament coasts in a country like Kenya and ways and means to simplify pharmacy administration.

d) Regional distribution of health services. The effectiveness and accessibility of medical services depend intrinsically on their distribution among the population and within the area of settlement.

In turn the distribution of population is the result of a long development based on natural conditions, such as precipitation, soil fertility, and the consequences following from them, including the administrative division of a country. To begin with, the comparative valuation of particular districts as far as the number of inhabitants per medical service unit or physician is concerned reflects only the overall developmental situation of a district or a region rather than the state of health service facilities in a selective sense. For an evaluation of these, an investigation of hypothetically evenly distributed medical services would make sense only in comparable regions having a similar state of development. On the other hand, it is pointless to prove that underpopulated regions are also medically "under-provided". The fact of "under-population" or rather low population density is the result of given conditions over a period of time, together with the

development potential which does or does not result from it.

The present distribution of population in Kenya, as well as the size of its administrative units, have therefore to be regarded as a development parameter and a parameter for the ecological endowment of an area.

A distinct relationship between the population figure and the district area exists in the sense of a decreasing density of population with increasing area of district (see Map. 6) and results in four possible sets of relationship:

a) districts with a large population total in a small administrative area, b) districts with a large population total in a large area, c) a small population total in a large area, d) a small population total in a small area.

Undoubtedly the economically active areas with a high agricultural capacity and corresponding potential for development are also those with a high density of population, whereas sparsely populated or overloaded areas constitute marginal zones in respect to development policy (see Chapter A II).

An evaluative analysis of the health services has to draw comparisons within comparable areas and to investigate areas not open to comparison as to their differences on the basis of the population-area relationship. The investigation which follows will therefore arrange the districts in four comparable groups. Instead of making a comparison at national level, the health services will be studied within these four groups. Such a semi-quantitative evaluation of each district (relating to the number of inhabitants and the area to be covered per unit of medical services) permits a rather more detailed presentation of the regional distribution of medical services. For this purpose all the relevant data obtainable at district level were related to area and population number and entered in a scoring system. Even this, however, did not provide sufficient emphasis on the often very unhomogeneous distribution, even where it occurs within some extensive districts with a very thin and uneven distribution of population. But for methodological reasons it was necessary to abide by district boundaries.

The population or areal relationships arrived at for each medical provision parameter are arranged in 6 quantitative classes (6 – 1) in descending order. The class division, which refers to the relationship of population to medical provisions, is related to the situation in Kenya and not, as might be assumed, to the WHO recommendations for developing countries.

In quite a number of districts a more than an optimal relationship to the number of inhabitants arises, whereas in others relating it to the population an optimal provision appears to exist if the district's population density is ignored. To avoid this, the relationship to area was defined by the radius of a hypothetical catchment area (Diesfeld, 1970), on the assumption that even for hospitals outside a radius of 30 km, i.e., a catchment area exceeding 2800 square km, the attendance chance is practically nil for the population living beyond and is therefore statistically insignificant as well.

Table 13. Population and area ration per hospital in 41 districts.

Number of inhabitants/ hospital	<25,000	25,000 –	50,000 –	75,000 –	100,000 –	150,000 and more
No. of districts	5	6	8	10	4	8=41
Score	6	5	4	3	2	1
Radius (r) and	<10 km	10 – 19 km	20 – 29 km	30 – 39 km	40 – 49 km	50+ km
Area (A)/hospital	314 km²	314 km²	1,256 km²	2,826 km²	5,024 km²	≫7,850 km²
Number of districts	3	9	9	4	3	13=41
Score	6	5	4	3	2	1

Particularly in the large districts, the area as a reference figure becomes a valuable aid in the assessment of provisions of medical services. On the other hand, due to the uneven distribution of population, it is another instance of the methodological drawback of employing an administrative reference area as opposed to the truly functional reference area of the defined catchment (Diesfeld, 1970).

The following parameters of medical services * at the district level were related to area and population on the basis of the 1969 Census:

aa) Hospitals: government hospitals, mission hospitals, and other non-government hospitals with more than 20 beds. As a rule the small private hospitals in the major towns, especially those of Nairobi and Mombasa, were not included since their significance for the medical care of the population as a whole was restricted. Hospitals of 20 beds or less and open to the general public are listed in the health centre category, especially since they are mostly run without a physician. Special hospitals, such as mental hospitals, prison and military hospitals, and private nursing homes, are not included (Table 13).

bb) Hospital beds: As an indicator of the capacity of these hospitals the number of beds in all government, mission, and other public hospitals with more than 20 beds in any one district was compared with the number of inhabitants (number of inhabitants per hospital bed) (Table 14).

cc) Physicians: The number of government doctors per district was assessed on the basis of the most recent information from the Ministry of Health (1974); the large number of doctors at medical centres such as the Kenyatta National Hospital or the Provincial General Hospital was attributed to the respective district in which the main regional hospital is situated, on the grounds that it is only the population of the immediate vicinity that benefits significantly from the presence of such personnel.

The state of the register of non-government doctors is so inaccurate that they were not included. This is of no great importance in the case of the few doctors in

private practice in the larger towns, as their effect on the overall population is small, but it does lead to a measure of undervaluation of mission doctors. Reckoned at district level, the distribution of population per hospital bed and government doctor is as follows:

Table 14. Hospital bed and doctor ratio per population in 41 districts

1 bed/ population	Dis- tricts	Score	1 doctor/ population	Dis- tricts	Score
<1,000	17	6	<10,000	2	6
			10,000 and more	6	5
			20,000 and more	4	4
1,000 – 1,999	10	4	40,000 and more	9	3
>2,000	14	2	80,000 and more	11	2
			160,000 and more	9	1
	41			41	

dd) Health centres: Mission hospitals with 20 beds or less are classified with government health centres; their population and area ratio was assessed for each district and resulted in the following distribution and score (Table 15).

International comparisons would suggest that the optimal population relationship lies at 25,000 – 50,000 inhabitants per health centre. The radius of access to a health centre, which a patient might be expected to negotiate, is about 15 km, and the catchment area is accordingly a maximum of 700 square km.

Dispensaries: All the dispensaries listed in official statistics—whether maintained by central or local government, by the municipalities or missions, or privately run by plantation companies or industries,—have been assessed for each district and related to population and area (Table 16).

This shows that two-thirds of all dispensaries are required to serve more than 15,000 inhabitants and a radius of over 9 km. In one-third of all districts the area to be served by one dispensary is in excess of 1200 square km.

Although population density constitutes a very important variable in the capacity of a medical system, the extremely unhomogenous distribution of population in Kenya necessitates a separate investigation of area and population total, albeit with the latter arranged in three groups according to population density:

---

* The data upon which the scores are based as well as the calculations and original tabulations from which the scores are derived are in the possession of the author (Diesfeld) and also deposited at the Geomedical Research Unit of the Heidelberg Academy of Sciences.

Table 15. Inhabitants, radius and area per health centre (H.C.) in 41 districts

Inhabitants/ H.C.	<25,000	25,000 –	50,000 –	75,000 –	100,000 –	≧ 150,000 or no H.C.
Number of districts	3	6	5	10	6	11
Score	6	5	4	3	2	1
Radius (r) km and	<10	10 –	15 –	20 –	30 –	40 and more
Area (A) km²/H.C.	<314	314 –	706 –	1,256 –	2,826 –	5,024 and more
Number of districts	3	5	8	5	5	15
Score	6	5	4	3	2	1

(a) districts with a population density of up to 10 inhabitants/km²

(b) districts with a population density of 10 – 100 inhabitants/km²

(c) districts with a population density of 100 – 1,000 inhabitants/km²

Following analysis of the statistical data (see Table 16) a median score for medical service per district was arrived at in accordance with their overall distribution; this was computed separately for *area* (km² per hospital, health centre, dispensary) and for *population* (inhabitants per hospital, hospital bed, government doctor, health centre, dispensary). As had been anticipated, in areas with a population density of less than 10 inhabitants per km² and with a large area, what appears as a relatively favourable service provision in respect to population numbers is usually cancelled out by the sparse settlement of large administrative areas. This fact becomes quantifiable when referring to the relatively large difference between *"area score"* and *"population score"* (see Map 6a). If the population continues to concentrate around particular central places, a relatively favourable situation may come about in spite of the extensiveness of the area (e.g. Marsabit, Isiolo, Tana River, Lamu, Taita). The effectiveness of medical services in the districts increases quantitatively and spatially in line with increasing population and decreasing area

A relatively favourable service density "population score" occurs in comparison to a relatively unfavourable "area score" above all when there is an unhomogeneous distribution of population — as happens in Machakos and Kitui districts. A slight difference in the population and area service density in these areas with a population density below 100 inhabitants per km² indicates a relatively good quantitative and spatial degree in the effectiveness of medical provisions.

Areas with high densities of population, especially in western Kenya, reach the point at which, relatively favourable spatial conditions notwithstanding, the high numbers of population cause medical services to fall quantitatively below standard (Kisii, Kisumu, Siaya).

Despite high population figures, the Central Highlands still enjoy relatively good service (Nyeri, Embu, Kiambu, even Meru), whereas Murang'a appears to be inadequate with regard to population figures in spite of a favourable spatial relationship. Both the urban centres of Nairobi and Mombasa present favourable conditions in this respect (see Map 6). In summary, it may be noted that notwithstanding the definite limitations resulting from an insufficiency of information, this analysis of medical service density at district level relative to population and area does succeed in demonstrating the core areas and bottlenecks of the services in somewhat greater detail. It thus highlights what areas are in the final instance problem areas in development policy in the field of health services.

6. Medical Manpower and Training of Medical Personnel

For a long time Kenya placed its staffing emphasis on physicians, a policy in accordance with the primarily curative and hospital-oriented nature of the health services. On the other hand, it was obvious that if the system persisted in this manner, the demand for doctors could not be met in the foreseeable future, even if Kenya were to open its own medical school. Even if Kenya — with a global doctor/population ratio of 1 : 13,000 at the present time — comes close to the WHO recommendation of a ratio of 1 : 10,000, thus topping the league of physician density in Africa, it still suffers from gross regional unevenness in their distribution. Then as now the mass of the people over large parts of the country has

Table 16. Inhabitants, radius and area per dispensary in 41 districts

1 Dispensary/ inhabitants	<10,000	10,000 –	15,000 –	20,000 –	25,000 –	30,000 and more
Number of districts	4	9	9	11	4	4
Score	6	5	4	3	2	1
1 Dispensary radius (r) km	<6	6 – 9	9 – 12	12 – 15	15 – 20	>20
and area (A) km²	<100	254 –	452 –	706 –	1,200 –	1,200 and more
Number of districts	3	10	5	7	0	16
Score	6	5	4	3	2	1

little chance to receive medical attention from a physician when needed.

The idea of employing paramedical personnel for basic medical care in Kenya dates back to World War I (see Ch. B. 1). During the twenties, medical assistants became an integral part of medical care, the accent being on "auxiliaries" and "medical assistants" for the doctors. More modern aspects of paramedical activities, given the appropriate training, emanated from East Africa, and in particular from Kenya, and pointed the way to further developments in the future (Fendall, 1963; King, 1966, op. cit.).

Within the framework of the Kenyan health services, paramedical personnel fulfill an important role in the field of the basic health services. Separated from the specialised hospital services, these are headed by physicians who regard themselves in a new role as instructor, organiser, planner, and administrator.

In the past, training policies were formulated on the basis of investigations into demands carried out by the ministry with the aid of the WHO. But more exact data were not then available. It was only in the early seventies that working parties from the Ministry of Health, the WHO and the Department of Community Medicine of the newly established Medical School, together with the Dutch research team led by Dissevelt and Vogel (1971), Dissevelt (1972), Vogel et al. (1974) of the Medical Research Centre of the Royal Dutch Tropical Institute, Amsterdam, were able to supply more precise data from their survey of the functions and effectiveness of rural health units, on the basis of which long-term projections of demand could be made as a foundation for future planning.

a) The medical profession: In 1960 there were 700 and in 1973 more than 800 registered and licensed medical practitioners in Kenya. Most of these were non-Africans, that is of British, Asian or other non-African origin. The data available, for example in the East African Medical and Dental Directory 1973, are not exact because although new arrivals have been entered, departures have not been correctly deleted. As a result the actual number of practicing doctors can never be accurately gained from it, and accordingly the figures in Table 17 must be regarded as considerably inflated.

On the other hand, the Development Plan 1970 – 74 makes the assumption that there were 762 actively practicing physicians in the year 1968 (op. cit. p. 497).

Table 17. Registered Medical Personnel

Medical Practitioner	1960	1969	1970	1971	1972
Registered	713	1090	1218	1405	1601
Licensed	53	213	306	381	481
Total	766	1303	1524	1786	2081

Source: Economic Survey 1973, Central Bureau of Statistics, Ministry of Finance and Planning, Nairobi, Nairobi 1973, p. 193

Table 18. Active doctors by type of employment and location, Kenya 1967 (Wheeler, 1969)

Type of employment	Location			
	Nairobi	Mombasa Nakuru Kisumu	Rest of Kenya	Kenya total
Central Government	100	30**	136	266
Local Government	33	–	–	33
Industries	–	–	6	6
Missions	–	–	60	60
Private practice	247	105	62	414
Total	380	135	204	779
Population 1967 *	438,000	307,000	9,203,000	9,948,000
Doctor/Population ratio	1 : 1,153	1 : 2,274	1 : 34,860	1 : 12,700

* 1967 population figures = 1962 census figures plus 5% p.a. for urban areas, 3% p.a. for total population
** Estimate based on 10 doctors per provincial general hospital — compiled from Annual Reports, Ministry of Health

Data reproduced there are largely identical to those figures employed by Wheeler (1969) in a more far-reaching analysis, on the basis of which he projects the demand for physicians to the year 2,000 (see Table 18).

Calculating on the basis of the 1962 Census, a population of almost 10 million, Wheeler arrived at a physician/population ratio of nearly 1 : 13,000, although he somewhat blurs the dichotomy between town and country. For Nairobi and the urban centres of Mombasa, Nakuru, and Kisumu, the physician/population ratio is 1 : 1,500. Almost half of the physicians in Kenya reside in Nairobi, although only 5% of the country's total population lives there; in consequence the physician/population ratio falls to 1 : 1,153 here. Only one-seventh of all doctors in private medical practice lives outside the four towns, and even these probably practice in the main district towns and are chiefly available to the non-African population as in the urban centres mentioned above. The actual ratio of physician to population for the African population living in rural areas of 1 : 40,000 is certainly no understatement, and it is probably even further aggravated by the uneven distribution in line with the varying degrees of development in rural areas: The "personal emoluments" of the Ministry of Health or their "estimates of recurrent expenditure" for 1973/74 show 453 doctors, including those in administrative services, 37 of whom were senior medical officers, 30 medical specialists, 14 senior medical specialists, and 472 Medical Officers I and II.

A further, very important aspect is the fact that as late as 1967 only 15% of the doctors were Africans (see Table 19). Even the manpower survey of 1972 only lists 36.9% (=308) of the 834 doctors as citizens. Non-citizens therefore constitute 63.1% of the total, and it is almost certain that a number of naturalised non-Africans are entered as citizens. (Kenya Statistical Digest, Vol. X, No. 4, Central Bureau of Statistics, Nairobi 1972, p. 7).

Table 19. Active doctors by citizenship and type of employment (Wheeler, 1969)

	Citizens	Non-Citizens	Total
Central Government	73 ⎫ 93	193	266
Local Government	20 ⎭	13	33
Missions	–	60	60
Industries	–	6	6
Private Practitioners	21	393	414
Total	114 = 14.6%	665 = 85.4%	779 = 100%

The present strength of the staff in medical service under the Ministry of Health, excluding the central health administration, amounts to 413 doctors, some of whom are delegated by voluntary agencies and therefore appear on the payrolls of the Ministry. The statement in Tabel X (Annex) also shows that 43% of all doctors in Kenya are employed at the Kenyatta National Hospital and that considerably more doctors work in the provincial general hospitals (37.8%) than there are in all the district hospitals combined of the province concerned (19.1%).

The number of non-citizens will decline over the coming years, and this too will have to be taken into account when planning the training capacity for the Kenyan medical staff. Moreover, when projecting the demand for physicians, losses through retirement on age grounds or through a "brain drain" should also be considered. Until 1972 the demand for doctors will have to be met by foreign university graduates.

Medical training: The Faculty of Medicine at the University of Nairobi, then the University of East Africa, was opened in 1967. Until then it had been possible to carry out only the bed-side teaching of the interns from the Medical School of Makerere University College, Kampala in Uganda. "The objective of the Faculty is the education of a doctor capable of organising and carrying out all the work required at the district level in Kenya. In achieving the objective the Department of Community Health plays a major role" (extract from "Information on Department of Community Health, University of Nairobi, January 1974). The number of students in the first year of study has risen continuously. In 1972 the first cohort of 16 students graduated with the M.B., B.S.; a year later graduates numbered 33. New admissions in the same year amounted to 105. In the academic year 1973/74, a total of 400 medical students had registered.

Wheeler's analysis (1969) was confirmed in its tendency by the number of graduates in 1972 and 1973, so that his prognosis that until the year 2000 the doctor/population ratio would undergo only insignificant fluctuations from the present one of 1 : 13,000 still holds good. His most important pronouncement was that unless there is a dramatic decline in the growth of population or a considerable increase in the number of doctors, the WHO target of one doctor per 10,000 people

will certainly not be reached by 1980, and probably not by 2000 either. He discerned the following implications in his projections:

"... Given doctor/patient ratios of the magnitude indicated above, it is clear that it will be impossible to restructure the pattern of medical care in Kenya radically (unless a much lower priority is given to hospital services than at present, which appears unlikely). Specifically, perpetuation of the present structure means that hospital doctors will continue to be aided by medical auxiliaries, and that the health centre and dispensary services will be staffed exclusively by auxiliaries.

From this conclusion, two further ones follow; firstly, doctors must be prepared for a situation in which they will be working alongside and supervising medical auxiliaries, where a large part of their work will be teaching and administration: for this they must be explicitly taught. Secondly, the output of medical assistants will have to be expanded at least to the level of output of doctors if there are to be sufficient of them to meet the needs of hospitals and health centres. Further, since supervision of scattered rural health units is difficult, there is a strong case for raising the quality of medical assistants so that they can take effective charge of the day-to-day running of the health centres ..." (Wheeler, 1969).

The consequences of this emerge in a study of the new plans for the training of paramedical personnel.

b) The paramedical personnel: Study of the staff categories and training programmes of paramedical personnel shows that as far as the immediate future is concerned, the expansion of the basic rural health services will remain problematic in spite of recent demands for increases — and this not only because of the long take-off time due to the expansion of the training programme and training capacity, but probably also for overall economic reasons.

The term "paramedical personnel" is differently defined by different countries and authors. It includes all medically trained personnel more or less employed in the running of a hospital in its nursing, technical and assistance aspects (e.g., registered nurses, pharmacists, laboratory technicians, x-ray technicians, etc.), as well as those who are unfortunately known as "auxiliaries", who are not only secondary hospital personnel but who, within the framework of the basic rural health service, are often left on their own with no more than irregular supervision by a doctor.

Building on the early experience with paramedical personnel in Kenya 60 years ago and based on the training courses set up for British nursing staff, corresponding training centres were established, first and foremost in the mission hospitals. These were followed additionally by ones in the Medical Training Centre (MTC). Nairobi, and in the provincial general hospitals (Table 20).

Table 20. Number of training institutions for paramedical personnel of enrolled level (Migue, 1964)

Personnel Category	Government	Church	Total
Enrolled nurses	8	13	21
Enrolled midwives	4	12	16
Public Health nurses	3	–	3
Psychiatric nurses	1	–	1

Source: M. R. Migue, Paramedical Education. In: Vogel, L. D. et al., edit. op. cit. p. 149

Registered nurses, registered midwives, medical assistants, clinical assistants, and other technical staff are being trained in Nairobi at various high-level hospitals.

Basic courses: Basic courses leading to the enrolled level are open to candidates with the certificate of primary education (CPE), the Kenya Junior Secondary Examination (KJSE), and to those who have passed Form IV or the East African Certificate of Education (EACE). Countinuing to build on these, in some subjects post-basic courses are offered, leading to Registered Nurse (KRN) or Registered Midwife (KRM).

A clinical officer (formerly a medical assistant) receives his training in two stages at the Medical Training Centre, as a one year extended training course of an experienced enrolled nurse leading to certified clinical officer or, on the basis of the EACE, by way of a three year course to registered clinical officer. The certified clinical officer grade is to be abolished. The course leading to registered clinical officer is to be expanded in order to meet both possible tasks: namely, to work in the curative capacity in the hospital under the guidance and supervision of a medical doctor, or to work as a general medical practitioner in place of a doctor as the team-leader in a rural health centre (King, 1966; Fendall, 1972). There is talk in Kenya of the clinical officer remaining the key figure in rural medical provision throughout the next century (Migue, 1974).

The projected demand for paramedical personnel, who form the rural health team in the health centre and in the dispensary, foreshadows an appropriate expansion of the training capacity initially until 1984 (Proposal for the Improvement of Rural Health Services, op. cit.). Besides the clinical officer, who is supervised and directed by the district medical officer, the rural health team consists ideally of an enrolled nurse, an enrolled community nurse, enrolled midwife, enrolled public health nurse, and a public health technician.

The intention is that the community nurse should be a multi-purpose worker who can double as a general nurse, midwife, and public health nurse. There is no room here to discuss the problems of the curriculum, refresher training, and the implementation of the programmes set out in the proposals.

The number of registered and enrolled nurses and midwives rose almost threefold over the period 1963 to 1972 (see Table 21). The projection for the next ten years up to 1984 plans a rate of increase in the important personnel categories, such as the clinical officer and the various enrolled nurses, twice as high as in the field of the hospital-based personnel. An analysis of the list of the existing paramedical personnel in 1971 shows the following urban/rural distribution and division according to employer and affiliation (Proposal for the Improvement of Rural Health Services, op. cit.).

Of the *Clinical Assistants* (n=687), 98.5% were citizens and 94% in government service, 81% were employed in places with more than 20,000 inhabitants, i.e., in

Table 21. Registered and enrolled nurses and midwives

	1963	1966	1969	1972
Registered nurses	1,824	2,332	3,099	4,141
Enrolled nurses	2,308	3,182	3,872	5,174
Total	4,132	5,514	6,971	9,315
Registered midwives	900	1,067	1,473	1,844
Enrolled midwives	871	1,217	1,584	2,237
Total	1,771	2,284	3,057	4,081

Source: The Kenya Gazette and Ministry of Health, 1974

Nairobi, Mombasa, Nakuru, or Kisumu, and only 19% in rural areas or district hospitals.

Among the *Registered Nurses* and *Registered Midwives* (n=1.053) there were still 27.5% non-nationals, the majority of whom were employed in mission hospitals. However, among the nationals 87% were in government services, 36% of them in rural areas.

The *Enrolled Nurses* (n=1,860) counted 99% of their number as nationals, 51% being in the rural areas, whereas among the *Enrolled Midwives,* nationals made up 96.5%, with 63% being employed in rural areas; about two-thirds of both groups were in government services, although about half of them had been trained in non-government hospitals.

One hundred per cent of the Health Inspectors (n=517) and Health Assistants (n=199) were nationals and in the government service; of these, 92.3% of the former were in large towns, and 44.7% of the latter in the rural areas.

This breakdown shows that especially the clinical assistants, who along with the KRN and KRM, are concentrated in the larger hospitals, fail to fulfill their intended role in the basic health services. Accordingly it is probably essentially the EN and EM who, if they do so at all, sustain the basic health services beyond the hospital.

One precondition for the improvement of this situation was created with the law providing for the take-over of the basic health services by the Ministry of Health.

7. Hospital Insurance Act and Social Security

a) The National Hospital Insurance Act: The National Hospital Insurance Act of the year 1966 in its revised form of 1967 (Laws of Kenya, Chapter 255, The National Hospital Insurance Act, Govt. Printer, Nairobi 1967), set up a National Hospital Insurance Fund under the control and management of the Minister of Health, who is assisted in this capacity by a National Hospital Insurance Advisory Council.

Under the terms of this act everybody with an income of Ksh 12,000 a year or more must contribute Ksh 20 per month. The insurance benefits consist of payment of daily hospital charges for up to a maximum of six months per year for the contributor, his wife, and his children up to the age of 18 (or beyond this if continuing in full-time education). These benefits are periodically reviewed.

In the case of employees, the employer is responsible for the proper deductions. Voluntary contributions can be made under a „National Hospital (Voluntary) Insurance Fund". The contributions, fixed at Ksh 10 a month are only half as high as those demanded from the compulsory contributors. Membership of the two funds can be altered in accordance with income.

The operating costs of the National Hospital Insurance Fund are provided by the budget of the Ministry of Health. In 1968/69 they amounted to K£ 46,000; in 1973/74 more than K£ 62,000 were earmarked. From 1968/69 – 1971/72 contributions rose to over K£ 1 million, amounting to an annual sum of about K£ 280,000 to K£ 300,000 (see Table 22).

Table 22. National Hospital Insurance Fund 1968/69 – 1971/72

Year	Receipts *	Benefits	Contributions net of benefits K£
1968/69	845,775	574,966	270,809
1969/70	964,666	622,786	341,880
1970/71	950,659	665,327	285,332
1971/72	1,026,576	777,635	248,941

Source: Republic of Kenya, Economic Survey 1973
* Central Bureau of Statistics, Ministry of Finance and Planning Nairobi 1973, p. 192

Besides the government hospitals, which are obliged not to exceed services beyond the maximum Ksh 35 per day in the hospital, a number of private nursing and maternity homes are also entitled to services. The rates of the former are Ksh 55 a day, and for maternity cases generally about Ksh 25 is charged for the first five days.

As an incentive for voluntary membership, the contributers are eligible for the same benefits as the compulsory members, in spite of their lower contributions. Since 1972 voluntary membership has been irrevocable.

Figures of membership were not available, but in order to gain an impression of the scale of persons insured in this manner an estimate of membership was arrived at by assuming that the total contributions recorded in the statistics for those years were rendered by members in the compulsory category who paid Ksh 240 a year. On this basis the membership was found to be 3,500 – 4,000 —a modest start for a National Hospital Insurance Fund and an indication of the very small class of those whose incomes rise above Ksh 12,000 a year. Accordingly, the

membership does not amount to more than 0.5‰ of the population. This branch of insurance can therefore scarcely assume any sociopolitical significance.

b) The National Social Security Fund Act: Under the National Social Security Fund Act of 1965, in its revised form of 1967 (Laws of Kenya, Chapter 258, The National Social Security Fund Act, Govt. Printer, Nairobi 1967), a social security fund was established under the control and management of the minister responsible for social security. He is assisted by a National Social Security Advisory Council, composed of 5 representatives each of the government, employers, and employees. This, however, is a provident fund, the amount saved being paid out in the event of the insured risk materialising, and is not a genuine social insurance independent of the sum accumulated. The standard contribution amounts to about 10% of the salary, up to an incomes limit of Ksh 800 per month. Half of this is retained from the employee's salary, and the other half is paid by the employer. The state contributes to the fund only in its capacity as an employer.

The benefits consist of age benefit, survivor's benefit, invalid's benefit, withdrawal benefit, and emigration grant. Compulsory members are found among all employed persons in the public service, excepting the noncivil services, and all workers and salaried staff in regular employment in town and country, including farm labourers other than those engaged in temporary jobs. Those salaried staff or civil servants who are insured under other state or state-recognised superannuation or pension schemes are also excepted. Members and their families receive free treatment as patients in the government hospitals. In 1973, 600,000 employees, i.e. 5% — or with families 15% — of the population, were insured under such schemes (Musiga, L.O. Problems of social protection in Kenya, Int. Soc. Sec. Review 4, 1974).

Risk pooling does not occur in this system. But that some savings capital is redistributed from areas with good work opportunities to those with poor, is shown by the number of insured persons working outside their home region being greater in certain provinces, as for example, Nyanza Province, than that of employees in the region itself. In the event of an insurance having to be paid out, the capital earned and accumulated elsewhere is transferred to the home region. For insured persons in the lowest income group, however, this kind of social insurance is expensive, since it reduces the already inadequate income without providing real help. This is particularly the case under present rates of inflation.

C. The Diseases of the Country

Introduction: Source, Material, and Methods

Certain methodological problems are likely to be encountered if observed disease occurrences based on hospital or other medical statistics are interpreted in a spatial geographical analysis. Epidemiological surveys of good quality — taking environmental factors into proper consideration — are scarce and usually restricted to a few diseases of major interest or importance and consequently generally very specific as opposed to aiming at spatial coverage.

The most prominent problem is the lack of a *common spatial denominator*. For example, in order to examine the relationship between rainfall or temperature and disease, the two physical variables are demarcated by isohyets and isotherms, whereas disease prevalence available in terms of hospital records and medical statistics for notifiable diseases may be based on administrative units.

Hospital statistics or other health statistics suffer greatly from a number of systematic and mathematical biases, which can scarcely be overcome. Some epidemiologists consider them to be useless, but since there is usually no better information available they will have to be used for a long time to come, albeit with great care. Two sources of data are available and can be compared with existing epidemiological surveys and government reports on the disease distribution.

The most commonly used information is derived from *weekly notifications of infectious diseases* on a district basis, and set out in relation to district populations although very little can be said about the distribution within the district and the completeness of this information.

In English-speaking countries such as Kenya, *annual returns* from hospitals give a rather dense but dotted, non-contiguous coverage of a number of diseases according to the *International Classification of Diseases* (ICD).

By expressing the median number of diagnoses per year entered in a hospital record over a period of at least 8 – 10 years as a rate per 10,000 or 100,000 population, a Hospital Recording Rate for the hospital catchment area of, in the case of Kenya, 54 record-keeping Government hospitals becomes quite instructive about the spatial aspect of a disease and is open to careful interpretation.

$$\text{Hospital Recording Rate } \lambda = \frac{\Sigma\, x_i}{n_i}$$

The hospital recording rate is here defined as the frequency (λ) with which a certain disease (x_i) has been observed and recorded in a hospital (i) and entered in the annual return of diseases. It is assumed that the patients having these diseases came from a population (n_i) in general within the catchment area of the hospital. The population 'at risk' in this case and having a chance of

being seen, treated, and recorded in the hospital, is the denominator, to which the reported diseases are then related. The observed distribution frequency is best approximated by a Poisson distribution and was previously ranked in a logarithmic scale. Studies by Hinz (personal communication, 1974) showed that a stanine rank transformation, classifying the hospitals according to an ascending order of the Hospital Recording Rate for each disease in 9 ranks differentiated (the hospitals) somewhat better and without an unjustified impression of accuracy:

Stanine Ranking of Hospital Recording Rates

1st = 4%	
2nd = 7%	
3rd = 12%	
4th = 17%	of all recording hospitals according to an ascending
5th = 20%	order of disease recording rate per population in
6th = 17%	catchment area
7th = 12%	
8th = 7%	
9th = 4%	

The hospital catchment area has been defined for 54 Kenyan Government hospitals according to a method described earlier (Diesfeld, 1969 a), whereby the population within the catchment area was estimated according to the Kenya population census 1962 (see Fig. 4). The median number of cases of each diagnosis was taken as the statistical median value for the effective number of years, for which annual returns were available. In general, the reporting period was the 10 years from 1963 to 1972. The median year was thus 1967, for which the population in the catchment area was estimated as three-quarters of the population growth between the census of 1962 and 1969. This was done by taking into consideration the change of administrative boundaries between the two censuses. In all it was possible to use 455 hospital/years, for the purpose of analysing the disease pattern according to the Hospital Recording Rate.

The advantages and disadvantages of this method have been discussed several times elsewhere (Diesfeld, 1969 a, 1973, 1974 a).

This method at least gives some semi-quantitative indications regarding the spatial distribution of certain diseases as they have been observed in Government hospitals over a timespan of 10 years.

Hospital statistics in Kenya on a larger scale have never previously been analysed, therefore such analysis of hospital statistics may well yield additional information.

Statistics from weekly notifications of infectious diseases and from the annual returns of Government hospitals over the past decade thus serve if not as an ideal then as a valuable basis for reflections concerning the spatial distribution of diseases. These reflections will be considered in the context of the scientific literature and other reports on the distribution of diseases in Kenya.

I. Diseases Transmitted by Arthropods

1. Malaria

(ICD 084) *

"Malaria is a general term applied to a group of diseases caused by infection with specific protozoa of the genus *Plasmodium* and transmitted by the infected female of various species of Anopheles mosquitoes." **

The following forms of malaria are distinguished parasitologically, clinically, and according to therapy, prophylaxis, and prognosis: Malaria tropica, caused by *Plasmodium falciparum,* Malaria tertiana, caused by *Plasmodium vivax,* and Malaria quartana, caused by *Plasmodium malariae.*

Malaria used to be the most significant single cause of mortality in the tropics and sub-tropics, until the WHO inaugurated its malaria eradication campaigns. Although half of the population — predominantly in Asia, which was formerly at risk is now free from malaria, there are still about 360 million people living in malaria-infected areas, the majority of which are in Africa.

But even in Asia malaria is once more on the increase — an indication in itself of the complexity of the problem of malaria control (Bruce-Chwatt, 1977).

According to the climatically-controlled chances of survival and reproduction of the mosquito vectors, malaria is transmitted either seasonally or all the year round. During transmission-free periods, the human constitutes the natural reservoir of the plasmodia, from which new mosquito generations can infect themselves and thereby maintain the transmission cycle. Nothing save continuous re-infection throughout the year can lead to a certain degree of immunity. In areas of year-round transmission, malarial mortality is particularly high among infants and young children who have not yet built up immunity — a matter of time. In contrast, in areas of seasonal transmission, occurring mostly at the beginning and towards the end of the rainy season, persons exposed to it develop a lesser immunity. The result is that not only babies and small children but also adults frequently succumb to clinical malaria. Seasonal epidemics may then arise, which can in their turn lead to higher rates of morbidity and mortality and, at times of intensive agricultural activity, to economic losses for rural families.

The distribution of malaria is determined by *human-ecological factors* such as the state of immunity, age occupation, the types of housing, sleeping-provision and settlement, the forms of agriculture, migration and the population density; set against these are the factors bear-ing on the *ecology of the transmitting mosquito,* comprising species of anopheles and their life habit, the blood sucking and manner of depositing eggs, their flight range, duration of hibernation and life, as well as their susceptibility to plasmodia. For both the plasmodia and the vectors, environmental factors like climate, temperature, relative humidity, altitude, and water types play an important role. Different species of plasmodia possess differing epidemiological characteristics. The state of immunity of the human host, the natural or acquired tolerance through total or partial immunisation, as well as premunition (=immunity built up and maintained only in the presence of parasites) influence the epidemiological situation of malaria in a given area.

To date no systematic malaria eradication campaign has been conducted in Kenya. In some parts of the country, malaria is still an important cause of illness (see Fig. 17, back of Map 9).

On the other hand, Kenya has played a significant role in the history of research into malaria, for it was P.C.C. Garnham, when working in the Division of Vector-borne Diseases in Nairobi in 1947, who elucidated the liver cycle of the malaria parasite.

Malaria has existed since time immemorial in the humid-warm coastal areas bordering the Indian Ocean and Lake Victoria. Before the arrival of the Europeans, altitudes above 1,500 m appear to have been malaria-free, for example, malaria epidemics being mentioned in Nairobi only from 1902 onwards (Symes, 1940; De Mello, 1947). Population movements brought about by World War I — such as recruitment from malaria-free areas, followed by assignment to malaria-infected areas and demobilisation and return to the home regions — contributed to the dissemination of malaria. Areas in Nandi and Uasin Gishu (Sirikwa County), which are situated at higher altitude, provide a case in point. After World War I the colonial development of the area by rail and road construction, the farm and plantation economy with its related labour movements, and the general increase in population mobility, all facilitated the further spread of malaria (Matson, 1957; Robert, 1964, 1974).

Indeed, John L. Gilks, a critical observer of the health situation during the colonial period, already considered malaria a social disease, which would not be successfully overcome without raising the standard of living of the public at large (Medical Report for 1929, p. 17).

It is possible to present the contemporary epidemiological situation in Kenya only in rough outline because large-scale malariological investigations are not available.

Figures from the hospital reports and the "Weekly Notifications of Infectious Diseases" from the district medical officers provide an inaccurate picture, since the true reference figures are unknown, and the diagnoses usually remain parasitologically unconfirmed. The diagnosis of "clinical malaria" often covers a multitude of other febrile illnesses. There is even a tendency to diagnose malaria more frequently as the season of expected malaria wears on. In those areas in which malaria

---

  * This disease and the subsequent diseases in the following chapters are given the three-digit categories according to the Manual of the Internat. Statist. Classification 1975 Revision Vol. 1; WHO Geneva 1977.

 ** The definition of this disease and of the following diseases are cited after the list of the Council for International Organizations of Medical Sciences CIOMS: Communicable Diseases, Provisional International Nomenclature; Geneva 1973.

occurs frequently, parasitologically unconfirmed febrile conditions are naturally enough more often diagnosed as malaria than in areas where it is rare.

Over a period of time, original bias can thus simulate a pattern of distribution which will not necessarily stand up to parasitological examination. It is further promoted by the map of malaria transmission in the Atlas of Kenya, long since reprinted but apparently uncorrected, which takes its bearings essentially from contour lines. Thus it represents only the duration and to a certain degree the intensity of malaria transmission over the course of the year (see Map 7).

The most important sources of information concerning the distribution of malaria are the Reports of the Division of Vector-borne Diseases of the Ministry of Health, which in addition to the Nairobi headquarters maintains branches in several parts of the country. From these focal investigations into the distribution of malaria, control measures and even research are conducted.

Plasmodium falciparum is the most common agent of malaria; the other species of plasmodium also occur, but do not play a major role.

In Kenya the most important vectors of malaria are *Anopheles gambiae* and *Anopheles funestus*. *Anopheles gambiae* prefers temporarily sunlit, muddy pools, often resulting from human activities. The high temperatures generated in these pools, promote the rapid growth of the mosquitoes, often within five days. For this reason *Anopheles gambiae* generally occurs at the beginning of the rainy season. *Anopheles funestus* prefers clear and shallow water with vegetation, especially swamps rich in vegetation, quiet lakes or ponds fringed with vegetation, as well as clear streams. Areas of seasonal inundation serve as breeding grounds for both species, with *Anopheles funestus* enjoying a certain ecological advantage. In some places *Anopheles funestus* presents an alternating density, inversely proportional to precipitation, with the result that it alternates with *Anopheles gambiae,* thereby ensuring transmission all the year round (Fendall and Grounds, 1965 c; McCrae, 1968; Roberts, 1974).

In Kenya zones of varying malaria intensity can be distinguished. The first criterion is the duration of transmission in the course of a single year; this is determined by the seasonally-fluctuating density of the mosquito vector, which itself in turn is caused by the seasonal fluctuations of precipitation, temperature, and humidity. A rough distinction can be made between transmission throughout the year, as against transmission for 3 – 6 months of the year, and for less than 3 months a year. A fourth grade is found in the seasonal transmission occurring in the vicinity of streams in arid areas, and a fifth one is constituted by those malaria-free zones lying above 2,000 m, since the vector normally is no longer found there.

Another criterion is the age-specific spleen rate and the parasite rate in the blood of children aged 2 – 9. If an enlarged spleen or malaria parasites are found in the blood of more than 75% of children in this age group, the term holoendemic malaria is applied. A 74 – 50% rate

of infection or enlarged spleen is termed hyperendemic, and a rate of 49 – 10%, mesoendemic; below 10% the term hypoendemic is applied. Using these criteria, different zones are distinguished in Kenya where malaria occurs in a characteristic manner (Fig. 17, back of Map 9; Table 23 and Atlas of Kenya, 3rd Edition, 1970 Map No 47).

Table 23. Malaria Epidemiology by Type and Area (from Roberts, 1974)

Classification/degree	Spleen rate age 2 – 9	Area
1. Endemic		
(a) holoendemic	>75%	Coast Province, coastal area; Tana River, Kano Plains, Taveta,
(b) hyperendemic	50 – 74%	North Nyanza, Bungoma, Busia, Shimba Hills (Coast).
(c) mesoendemic	10 – 49%	Machakos, Kitui, Thika; parts of North Nyanza, Murang'a and Embu below 1,300 m.
(d) hypoendemic	<10%	Meru, Pokot, Samburu, Isiolo, Baringo.
2. Epidemic	Variable	Highland over 1,600 m with high rainfall and dry areas with exceptional rainfall: Masailand, Nandi, Kericho, Kisii, NFD, Eastern Kitui, Londiani, Elgeyo.
3. No transmission (sometimes anophelism without malaria)	None	At altitude over 2,000 m: Aberdares, M. Kenya, M. Elgon (forest, moorland, plateaux).

The malaria situation in the Coast Province: The coastal province, including the lower course of the Tana, is considered a holoendemic malaria region with transmission occurring within a period of 6 month or more (Roberts, 1974). The continuous investigations and selected controls in some key projects by the *Division of Vector-borne Diseases* (DVBD) of the Ministry of Health have succeeded in considerably improving the malaria situation without, however, in the least affecting the overall situation in the coastal area. Thus in the coastal settlements, especially Mombasa and Malindi and the important tourist-oriented seaside resorts malaria has receded markedly due to drainage and insecticides.

Indoor spraying since 1949 has been combined with fortnightly chemoprophylaxis among selected population groups since 1966 and has achieved the virtual eradication of the parasite among children under 15 in the Malindi urban district; there has also been a lowering of the parasite rate to 2% in the periurban district, whereas more than 10% of the children of the surrounding locations still have malaria parasites. Prior to the introduction of controls, the parasite rate stood at 48 – 75% (Ministry of Health Annual Report 1962; DVBD Annual Reports, Nairobi 1969 – 1973). In Faza (Lamu) the use of insecticides has likewise succeeded in lowering the parasite rate to about 3%. In Kilifi, Mombasa, and the coastal section between Mombasa and Msambweni malariometric investigations carried out in 1973 showed a parasite rate of 6 – 8% (DVBD Annual Report, Mombasa 1973).

For years considerable efforts have been made in the various resettlement and irrigation schemes to reduce what was originally an extremely high malaria infestation. In the Shimba Hills Resettlement Scheme, where 70% of all children had malaria parasites in 1959, fortnightly chloroquine medication of the more than 7,000 settlers, combined with the use of larvicides, succeeded in lowering the rate to less than 5% (DVBD Annual Reports, 1965, 1972, 1973). Because of its originally catastrophic malaria situation, the eventual success of this project had been in doubt.

The irrigation scheme at Hola (Galole) on the Tana River and the Ramisi sugar estates near the Tanzania border have been subject to larvicidal and chemoprophylactic treatment since 1963. Here too the parasite rate among the settlers could be kept at a low level, falling to 6% at Ramisi and even below 1% at Hola (DVBD Annual Reports, 1969 – 73).

Western Kenya (Western and Nyanza Province and Sirikwa County): Western Kenya, humid and warm in the vicinity of Lake Victoria Basin, at an altitude of 1,100 – 1,500 m above sea-level, is the second most important of the holo- and hyperendemic malaria zones in the country. When Lake Victoria was experiencing high water levels at the beginning of the sixties, flooding carried the intensity of malaria transmission from the coastal strip far into the interior of the country (Ministry of Health Annual Reports, 1960, 1962).

The DVBD branch in Kisumu and lately also that in Kimilili (Western Province) controls mainly the Kano Plains with their rice and sugar cane plantations. In Kisumu, where the larvicidal activities in the urban district are supported by the DVBD, occasional chloroquine prophylaxes are carried out in the peri-urban district, where in 1965 the parasite rate was still five times higher (2.4%) than in the urban district (0.5%). However, a study group of the WHO and the Medical Research Council recently noted far higher rates of infection (DVBD Annual Report, 1973).

In some locations in the Nyanza Province—in Vihiga, Khwisero, and in the Kisii District—the introduction of fishponds to increase the production of protein has also considerably increased the mosquito density, particularly that of *Anopheles funestus*. More than 80% of the thousands of newly constructed fishponds were found to be infected, and this caused a considerable increase in the chance of malaria transmission and a correspondingly increased malaria morbidity (DVBD, Annual Reports 1965, 1973).

In *Nandi District* (Chemasse, Nandi Hills) and in the Uasin Gishu District (Turbo, Kipkaren, and Lugari Settlement Scheme) at altitudes between 1,500 and 2,000 m a very unstable malaria epidemic had spread in the course of agricultural development in the early fifties, striking a non-immune population. The same holds true for the Londiani and Molo townships on the railroad in the direction of Kisumu at altitudes of up to 2,500 m. Insec-

ticides, in combination with chemotherapy or without it, were able to lower the originally high parasite rate to 8 – 10%, and in some places to below 3.5% (Roberts, 1956, 1964, 1974).

In *Kericho District* Garnham's first experiments with D.D.T. in 1946 were successful in a very unstable malaria situation with a parasite rate of 37%, practically stopping malaria transmission to the present day (Roberts, 1974).

In *Central Kenya,* in a malaria situation very much dependent on the particular altitude, the control measures of drainage and larvicides are carried out only in the larger settlements. The extensive irrigation installations at Mwea Tebere (Embu District) in a mesoendemic area are again successfully controlled by the DVBD.

In *Machakos District,* usually a mesoendemic malaria zone with transmission during 3 – 6 months of the year, malaria epidemics occurred in 1951 and in 1961; there was a renewed outbreak in 1972 after remarkably few outbreaks over the intervening years. It was possible to contain this outbreak by massive application of chloroquine (DVBD Annual Report, 1972).

Apart from a few exceptions at the valley openings to the west, near Kitale, the high altitudes (above 2,000 m) of the *Rift Valley* are free from malaria. In the southern Rift Valley, in the Kerio Valley, and further north in the semi-arid areas, malaria occurs only occasionally and epidemically along the stream courses. In the Perkerra Irrigation Scheme at Lake Baringo, the DVBD carried out control measures. In the semi-arid areas of the north and north east the chances of malaria transmission are rather reduced, because scanty precipitation and a high level of aridity do not encourage the vector populations. Following extraordinary rainfall there may be local malaria epidemics to the extent that a sufficiently large reservoir of plasmodia is available in humans. Such a situation is favoured by the increasing mobility of the population.

As Gilks stressed as early as 1929 (op. cit.), the control of malaria is a question of raising the standard of living of the majority of the population. This still applies even though highly effective chemotherapeutics and insecticides are now available. In spite of the increasing development of the basic health services, that represent the backbone of malaria control, systematic malaria control is not possible without massive application of logistical, organisational and financial means, and includes above all the necessary cooperation of the entire population. Today, almost 50 years after Gilk's statement, these prerequisites are still not present in sufficient degree, although the population's preparedness to prevent malaria is clearly on the increase.

2. Trypanosomiasis
(Sleeping Sickness, ICD 086)

Trypanosomiasis in tropical Africa is an infectious disease found and caused by haemoflagellates of the genus *Trypanosoma* and transmitted by *Glossina* species (tsetse flies).

Parasitologically, epidemiologically and clinically two types of sleeping sickness occur in Kenya; Gambian trypanosomiasis caused by *T. gambiense,* and Rhodesian trypanosomiasis, caused by *T. rhodesiense.*

The vector of the former are tsetse flies of the palpalis group, the latter of the morsitans group, each having its own distinct biology, which thus influences the epidemiological pattern of the disease. In recent years *Glossina palpalis fuscipes* has also been found to transmit *Trypanosoma rhodesiense.*

The history of the distribution of human trypanosomiasis in Kenya and its present extent cannot be discussed without a short reference to the larger framework of sleeping sickness in East Africa.

The agent of sleeping sickness, *Trypanosoma gambiense,* which first spread epidemically in East Africa, was probably imported from the Congo Basin and the southern Sudan to Uganda in the course of increasing social, economic, and military mobility at the end of the nineteenth and the beginning of the twentieth century. The 1889 expedition of H. M. Stanley in search of Emin Pasha certainly played no more than a symptomatic role (Langlands, 1967).

The vectors of the *Glossina palpalis* group of tsetse flies were densely dispersed in the gallery forests of Uganda and Kenya, along streams, and in the coastal areas of Lake Victoria, so that the imported trypanosomes met an ample reservoir of vectors and thus with conditions favourable for the distribution of sleeping sickness among the population.

Trypanosomiasis

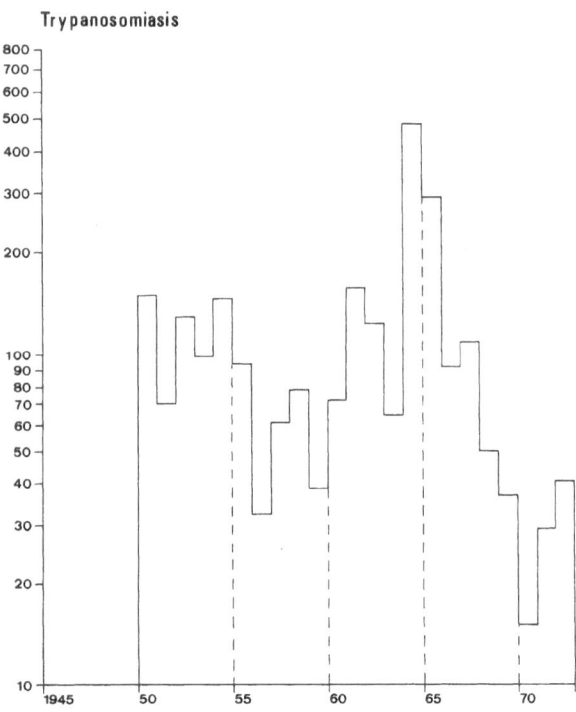

Figure 5. Trypanosomiasis, number of annual notifications 1950 – 1972 (Source: Willett 1965, Annual Reports EATRO, Bulletin of Infectious Diseases, Ministry of Health, Nairobi, 1962 – 1972)

Since the start of the twentieth century sleeping sickness has spread in several waves from Uganda to Kenya. The islands of Musenga and Mageta were the first to be affected, and from them it moved over to central and south Nyanza on the mainland (Onyango, 1974; see Fig. 15, back of Map 7).

In the course of the ensuing sixty years, Kenya experienced several epidemics, the first one of which lasted from 1902 to 1908. Emigration of the population allowed the disease to die down, so that a relatively low grade of endemicity was attained between 1911 and 1920. The deserted areas were left to revert to a natural state, increasing the density of Glossinae in the process. After 1920 the population increasingly returned to the deserted areas, thus enabling a second epidemic wave to rise between 1920 and 1930. However, at this time chemotherapy and bush clearance were already available as control measures, and the epidemic could be contained within certain limits. In a third dissemination wave from 1940 to 1950 the three districts of Kadimo, Alego, and Samia in the Western Province were particularly affected, together with the Nyando Basin west of Kisumu and the Kuju-Migori river basin in southern Nyanza. Massive control measures—above all the spraying of DDT in the gallery forests along the rivers—proved successful in suppressing sleeping sickness in these areas during the early fifties (see Map 7).

Until then it had largely been sleeping sickness caused by *Trypanosoma gambiense* and transmitted by *Glossina palpalis fuscipes.* Since 1940, however, Tanzanian migrant workers moving to the sugar cane plantations of Busoga in Uganda have brought with them *Trypanosoma rhodesiense,* which in turn encountered the vector *Glossina pallidipes* in these coastal areas which had reverted to a natural state, following the exodus of the population. Here this species had spread considerably, along with *Glossina palpalis fuscipes.* During the following decade and a half, particularly since the return of the population to the formerly deserted areas following the independence of Uganda, sleeping sickness once again increased. On this occasion it was caused by *Trypanosoma rhodesiense.* In the course of this spread a renewed importation of the sleeping sickness occurred from the west in Kenya.

Lesser outbreaks of *Trypanosoma rhodesiense* sleeping sickness, transmitted by *G. pallidipes,* occurred in the Sakwa and Mutonga Districts in 1953/54, in Samia in 1957/58, and in Yimbo in 1961. In 1958 Heisch et al. were able to identify the bushbuck as the animal reservoir. The rise in the surface-level of Lake Victoria caused by several years of abundant precipitation and the subsequent increase in humidity far into the hinterland led to a great increase in the Glossina population, again especially among *Glossina palpalis fuscipes.*

The years of 1964/65 saw an explosive outbreak (Fig. 5) in the Alego District with a total of 650 known cases. For the first time in the history of trypanosomiasis, the vector of *T. rhodesiense* was *Glossina palpalis fuscipes;* under favourable climatic conditions this has peridomestically spread from the immediate vicinity of streams and shore into the farmyard hedges and the thickets surrounding the fields and become very dense in numbers (Willett, 1965). This has demonstrated and confirmed certain earlier assumptions that *Glossina palpalis fuscipes* is able to transmit trypanosomae of the *T. gambiense* type as well as those of the *T. rhodesiense* type. In addition it proved possible to establish that the different transmission cycle by way of the *G. morsitans* or the *G. palpalis* group, and their respective preference for a host, already determines the virulence of a *T. rhodesiense* course or a *T. gambiense* course (Willett, 1965). Beyond this the assumption was confirmed that not only wild ungulates but also Zebu cattle may serve as potential reservoirs for the agents of trypanosomae pathogenic to humans (Onyango et al., 1966).

In southern Nyanza, the second focus in Kenya, the last outbreaks of the *T. gambiense,* transmitted by *G. palpalis fuscipes,* had been terminated in the Kuja-Migori area in 1958/59, due to very intensive anti-*glossina* measures on land and water (see Map 7).

Since 1962, however, there have been epidemic occurrences of sleeping sickness of the *T. rhodesiense* type over a limited area of about 500 farmsteads in the Lambwe Valley resettlement areas. In this fertile grassland *G. pallidipes* of the morsitans group was identified as the vector. A zoonosis was evidently encroaching upon a population recently immigrating within the framework of an agricultural resettlement project.

This focus is still not under control, whereas the old areas of sleeping sickness in the Western Province have largely been brought under control by application of insecticides, combined with bush-clearance, and planned agricultural utilisation with the close cooperation between the Division of Insect-borne Diseases of the Ministry of Health, the Tsetse Control Division of the Veterinary Department of the Ministry of Agriculture, and the recent support of the United Nations Development Programme (R. J. Onyango, 1974). The development of sleeping sickness in Kenya, particularly over the past 20 years, has once more demonstrated first, that new outbreaks of sleeping sickness arising from ecological shifts on the part of nature or man can occur at any time in areas long known to be infested by *glossinae,* and secondly, that these factors warrant special attention within the framework of agricultural and animal husbandry development projects. Correct planning with regard to certain ecological rules can arrest the danger, whereas wrong planning only conjures it up again. Although the tsetse fly, and especially the potential vectors of sleeping sickness, also occur over large parts of Kenya (see Tsetse Distribution Map, Kenya National Atlas, 3rd Edition, 1970), sleeping sickness has never left its original foci in central and southern Nyanza, a fact surely largely attributable to decades of efforts resulting in their control.

Figure 6. Kala azar, number of annual notifications 1945–1972 (Source: Fendall and Grounds, 1965 c, Bulletin of Infectious Diseases, Ministry of Health, Nairobi, 1962–1972)

3. Leishmaniasis
(ICD 085)

Leishmaniasis is a group of infectious diseases, which occurs in many countries and is caused by protozoa of the genus *Leishmania* and transmitted by sandflies *(Plebotomus spp.).* Of the three distinct clinical conditions only visceral Leishmaniasis (kala azar) occurs in Kenya; diffuse cutaneous Leishmaniasis has so far been found only in a few locations.

a) Visceral leishmaniasis (kala azar): An acute or chronic infection caused by *Leishmania donovani* and transmitted in Kenya by sandflies of the Synphlebotomus group. The clinical manifestation includes irregular fever, anaemia, leucopenia, splenomegaly, hyperglobulinaemia, finally, kachexia, resulting not infrequently in death.

Before World War II visceral leishmaniasis was a disease only sporadically observed among the nomads of the north. During the British Army's Ethiopian Campaign, between 1941 and 1943, an outbreak of kala azar occurred in the vicinity of Lake Rudolph which affected 135 soldiers (Fendall, 1952). Wakamba troops from the northern part of the Kitui District were probably responsible for its introduction at home, the first cases occurring there in 1948. Due to conditions especially favourable to transmission, a serious epidemic developed between 1952 and 1954, which spread further south along the eastern border of Kitui District and north among the Tharaka on both sides of the Tana river.

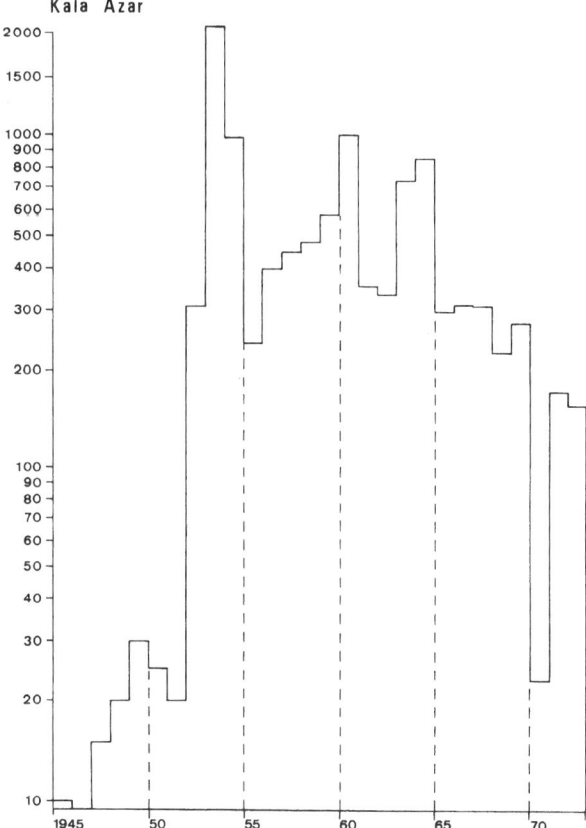

Kala Azar

The figures reported in the period 1954 – 1964 are essentially due to the development of this epidemic (Fig. 6). The ensuing slow decline in numbers is indicative of the self-adjusting endemic state.

The regional distribution of cases of kala azar is reflected in the analysis of Annual Returns of Diseases for the years 1968 – 1972; the following average values per 100,000 inhabitants in the catchment area of the hospitals are detailed below:

Marsabit	54	Tambach	12	Meru	6
Kabarnet	49	Lodwar	11	Kitale	6
Kapenguria	41	Machakos	9	Nakuru	6
Kitui	16	Garissa	7	Embu	3

Besides this relatively recent area of sedentary population, where the infectious condition has stabilised in the meantime in the sense of becoming endemic, investigations undertaken on account of these results showed further endemic foci of kala azar in the northern offshoots of the Rift Valley, in the Kerio Valley (Elgeyo Marakwet), in the Baringo and Samburu District, and in the West Pokot and even among the semi-nomadic herding population up to Uganda (Amudat) as well (McKinnon and Fendall 1955, McKinnon, 1962 a; Fendall 1961; Wykoff et al. 1968, 1969).

A recent investigation was carried out with the aid of the Montenegro skin test using the *Leishmania tropica* culture antigen in the well-known endemic area of the North Eastern Province with positive results in the three divisions of Tseikuru (29.5%), Gatunga (17.8%), and Nuu (20.2%). There was no positive case among 97 persons tested in the northern Rift Valley and only 5 positive outcomes among 116 persons tested in West Pokot (Mutinga and Ngoka, 1975).

Epidemiology: The epidemiology of the disease is essentially determined by the biology of the transmitting sand flies *(Phlebotomus martini)* and part of the Synphlebotom complex (Wijers and Minter, 1962, 1966; Wijers, 1963; Wijers and Mwargi, 1966). These sandflies exist in arid and semi-arid areas, mainly below 700 m, although they can still be found at an altitude of 1,000 m in West Pokot and Baringo. They hide in the ventilation shafts of *Macrotermes belicosus* anthills and only leave these on humid-warm, and calm evenings at the beginning and end of the rainy season. They have a very short flight radius. Indeed, it is only when man stops in the vicinity of anthills that he runs any risk of contact and transmission (Wijers, 1974). Here nomadic herding people come into sporadic contact when pitching their tents in the vicinity of the anthills. Although all age groups are affected, the greatest frequency occurs among children and younger adults (McKinnon, 1962 b). Among sedentary people those families dwelling near such anthills are the main ones to be affected; owing to the large number of anthills in this area, epidemics occurred between 1952 and 1954, in 1957, 1960 and again from 1963 to 1964. So far no animal reservoir has been discovered (Heisch, 1954, Heisch et al., 1956, Wijers, 1974).

As the endemicity increases, the peak of infections shifts to younger age groups. In the Kitui and Taraka areas, kala azar has by this time become a disease of young children (Southgate and Oriedo, 1962; Wijers, 1971). In some parts the disease has affected 70% of the population (Wijers, 1966). Wijers (1974) stresses that the infection rate is particularly high in hunger-stricken areas, among which the north of Kitui District and the Tharaka Division are to be counted, and that well-fed persons evidently appear to offer better resistance to this disease.

In some cases the disturbances in this area in 1965, which were connected with attacks by bands of robbers (Shifta), caused the population to leave the area, and the number of infections decreased, only to rise again after peaceful conditions were restored and the population returned to the area.

b) Cutaneous leishmaniasis, caused by *Leishmania tropica,* does not seem to occur in East Africa, although a few cases have been observed on the southern slopes of Mount Elgon in the Bungoma District since 1968 (Mutinga, 1975 a, b).

An investigation based on the Montenegro skin test showed a positive reaction among 16.5% of the 327 persons tested at Kapsokwony (Mutinga and Ngoka, 1975). Apart from several nonanthropophile phlebotomes, this new humid and relatively temperate area (in contrast to the hot and dry areas of occurrence of visceral leishmaniasis) was found to accommodate *P. pedifer,* which transmits cutaneous leishmaniasis in Ethiopia and the Sudan. It must be assumed that this species of phlebotome transmits cutaneous leishmaniasis here as well, but the experimental proof is still lacking.

Leishmaniases have been isolated from rock and tree hydraxes and a giant rat, and there is some suspicion that these animals may act as reservoirs (Mutinga, 1975 a, op. cit.).

The case of a cutaneous leishmaniasis contracted by a health worker trapping sandflies in the caves of the suspected animal reservoir has proved the transmission cycle recently (Arap Siongok and Birgen, 1976).

4. Arthropod-borne Virus Diseases
(ICD 060 – 066)

Arthropod-borne virus diseases (Arbovirus diseases) are a group of infectious diseases transmitted by arthropod vectors (mosquitoes). A number of arboviruses are known to be present or are assumed to occur or to have occurred temporarily in Kenya. Some viruses have been isolated, for others serological evidence was found (Metselaar, 1974 a).

Arbovirus infections are of public health interest in East Africa, where a very widespread epidemic of *Aedes aegypti*-transmitted chikungunya fever occurred in Tanzania in 1952 – 53 (Robinson, 1955), and was also observed in Kenya. A vast epidemic of o'nyong-nyong-fever, transmitted by *Anopheles spp.* raged through Uganda, Kenya, and Tanzania in 1959 – 1960 (Haddow et al., 1960). In both epidemics the infection rate was very high, yet fatal cases were not reported.

In a large multipurpose serological survey (1966 to 1968) in three selected districts: Malindi (Coast Province), Kitui (North Eastern Province), and Siaya (Nyanza Province), a total of 1,500 randomly selected sera were tested as to the presence of arbovirus antibodies (Geser et al., 1970). The sera were tested for antibodies against 3 Group A arboviruses (chikungunya, o'nyong nyong, sindbis), 6 Group B arboviruses (zika, yellow fever, West Nile, Banzi, Wesselsbron, dengue-1) and Bunyamwera virus. Chikungunya and o'nyong nyong infections, immunologically not distinguishable, showed the most dramatic geographical variation. Fifty per cent of all the people examined had antibodies in Siaya (Central Nyanza) and Malindi (Coast), whereas this antibody is virtually absent in Kitui, a fact perhaps explained by the dryness of this area and its low population density. The presence of antibodies even in the age group 0 – 4 indicates that this infection is still present in the population. This is supported by the observation of Bowen et al. (1973), who found that in the Nyanza Province south of Kisumu in an area on the coast of Lake Victoria proposed for an irrigation project, over 60% of schoolchildren had antibodies against the chikungunya -o'nyong nyong complex. The authors considered the irrigation project to be a severe threat to the population, as well as constituting a breeding place concentration.

Bunyamwera virus antibodies were isolated in 26% of samples from Malindi, only 5 – 6% of sera from Kitui and Siaya were positive. The same was observed for dengue-1 antibodies; 47% of the samples were found to be positive in Malindi. There is no clinical evidence of dengue fever. Metselaar, Henderson et al. (1974) have isolated from 147,000 mosquitoes in 7 areas of Kenya a number of viruses known to cause arbovirus infections in man as well. Among these were Semliki Forest virus (group A), Banzi virus (group B), Bunyamwera and Beliefe virus (Bunyamwera group).

As far as yellow fever is concerned, there was only one case in 1943 in Langata Forest in Kenya (Mahaffy et al., 1946). Kenya has never experienced yellow fever epidemics of the sort that have occurred in the neighbouring countries to the north—in the Nubian Mountains of the Sudan in 1940 and in south west Ethiopia between 1960 and 1962. In the course of years isolated cases of protective yellow fever antibodies have been found repeatedly in monkeys and bush babies. Similar investigations in Uganda have, on the other hand, disclosed that more than 30% of the monkeys are in possession of protective antibodies (Haddow, 1952).

Although Geser et al. (op. cit, 1970) in their sample survey found antibody reactions against yellow fever virus in more than half of the serum tests carried out in Malindi District, they were of the opinion that these were either cross-reactions to other arboviruses or the result of yellow fever vaccinations. However, the serious

epidemic in Ethiopia yielded proof of yellow fever protective antibodies in 7.5% – 15% of all human sera tested in Marsabit, Moyale, Lodwar, and Lokitaung in the northern districts of Kenya. The relatively high proportion found in Marsabit is attributed by Metselaar to the importation of yellow fever by members of the Ethiopian Burji tribe, who live there but now and then travel back to their home region, which was affected by the yellow fever epidemic in 1960 – 62. In the environs of the isolated, wooded sugar-loaf mountain of Marsabit, 150 km south of the Ethiopian border, neither protective antibodies among monkeys and bush babies nor yellow fever viruses among mosquitoes were to be found (Metselaar et al., 1970, 1974 b).

The problem of the potential transmitter of yellow fever is not unequivocally solved. In Kenya the classical transmitters *Aedes aegypti* and *Aedes simpsoni* do not appear to be particularly anthropophile. *Aedes africanus,* the vector of yellow fever, which lives in treetops, was not found in any of the three areas of Kenya in which yellow fever virus had evidently circulated some time ago. *Aedes simpsoni* was not found in the Langato Forest or in Marsabit, but only on the coast where it is not very partial to stinging humans; nevertheless, all three species of Aedes must remain under consideration as potential vectors of yellow fever, even in Kenya. According to Metselaar, the ecological balance may presently be such as to permit in principle the encroachment of yellow fever viruses wherever it seems to be present in primates, even if at present, according to all appearances, the probability is too restricted. Moreover, some aspects of the epidemiology and epizootology of yellow fever in Kenya are still not elucidated, so that constant vigilance and readiness are recommended to facilitate action in the event of any sudden requirement for yellow fever vaccinations.

5. Diseases Caused by Bacteria and Bacteria-like Organisms and Transmitted by Arthropods

This group of infectious diseases includes two important diseases or disease groups, in which rodents constitute an important natural reservoir for human infections.

For this reason these infections often disappear from the human context for years until they suddenly recur through contact with the animal reservoir. The occurrence of human cases is frequently preceded by unrecognised epizootics. In the case of plague, the observation of such an epizootic infection is of the greatest importance to prevent a threatening plague breaking out among the population.

Further typical representatives of this group are the endemic *tick-borne spotted fever* and the *endemic tick-borne relapsing fever,* which occupy a special position in so far as ticks act simultaneously as reservoir and vector, due to their characteristic ecological and biological behaviour. The epidemic infections of rickettsia and relapsing fever transmitted by lice make use of man himself as a reservoir and generally occur only in mass situations or catastrophies accompanied by widespread louse infestation. For years now these forms have scarcely been seen in East Africa and a description may therefore be omitted.

a) Plague (ICD 020): Plague, primarily an infectious disease of rodents, occurred in the past even in pandemic form, but outside its enzootic areas it is now reduced to mere foci in a few parts of the world. Large forest areas and highlands are ecologically predisposed to it. The agent *Yersinia pestis* is transmitted by rodent fleas such as *Xenopsylla cheopis* within the rodent population, as well as by the human flea *Pulex irritans*.

The natural reservoir exists in rodent populations, from which the agent is transmitted to semi-domestic and domestic rodent populations, and from these it passes to man.

Four clinical forms occur among humans: the bubonic plague, pneumonic plague, septicaemic plague, and abortive forms. Of the three pandemics during the Christian Era, the first Justinian plague of the sixth century is presumed to have affected East Africa (Pollitzer, 1954).

In 1697, during the Arab occupation, a plague epidemic occured in the then Portuguese Mombasa.

In the course of the third pandemic of the 19th and 20th centuries, the plague probably spread on the then customary trade routes across the entire African continent. During the closing decades of the 19th century, the various Christian missions thus repeatedly reported outbreaks of plague in East Africa. In Kenya it was mainly the construction and running of the Mombasa to Kisumu railway that played the important role.

Massive outbreaks occurred between 1900 and 1914, particularly in the vicinity of larger settlements (Nairobi 1902, Nakuru and Kisumu 1904, Mombasa, 1912). Nairobi experienced numerous outbreaks, the last one in 1941 with 781 cases (Roberts, 1950). In the year 1912, a plague epidemic occurred on the southern outliers of Kilimanjaro, which was described in great detail by Lurtz (1913), who already suspected at that time that it was not a matter of importation from outside via the domestic rat or by humans infected by plague, as was commonly assumed, but of encroachment from a local plague focus among the wild rodent population. As a result of the outbreak of war, this important observation was lost and only revived forty years later by Heisch et al. (1953).

From 1920 onward the rural foci of plague moved more to the fore: Rongai and Solai, north of Nakuru in the Rift Valley on the outliers of the Aberdares and Mount Kenya (Kiambu, Murang'a (Fort Hall), Nyeri, Kerugoya). No plague occurred there before 1914, whereas between 1921 and 1930 more than 2,000 cases were reported annually.

After World War II plague occurrences receded rapidly. A remarkable outbreak at Rongai in 1953 was carefully investigated and once again diagnosed as having been caused by wild rodents such as *Avicanthus niloticus* providing the reservoir for *Yersinia pestis* (Heisch et al.,

op. cit., 1953). In Kenya the last small outbreak of five cases and one death took place near Murang'a in 1963. A number of smaller outbreaks in Tanzania near the Kenya border, at Mulu on the Kilimanjaro foothills in 1968 (Msangi, 1969), and near Arusha (1970 and 1972) also indicate the presence of enzootic sylvatic plague, which can at any time encroach upon the human population, a possibility underlined by the investigation of Davis et al. (1968).

Applying the microhaemagglutination test to over 20,000 rodents of different species from different parts of the country, they discovered in 2.5 – 10% of the sera significantly high antibodies against fraction I of the kapsel antigen of the plague agent *Yersinia pestis*.

In its enzootic and epizootic course, plague is a typical example of a nesting disease, with a constant latent threat of encroaching upon human populations. Particularly the geography of Kenya with its secondary mountain ranges, the intermediate altitudes of the great mountain massifs, and the saddle of the Rift Valley are peculiarly conducive to the flourishing of rodents. Precise observation of the migratory movements and mortality of wild and semi-domestic rodents and rats in the urban districts is necessary and is carried out by the health authorities in order to recognise a sylvatic rodent epizootic in time to prevent the flood- or drought-conditioned transfer of a sylvatic rodent epizootic to an area which happens to be settled by humans.

b) Rickettsioses (ICD 080 – 083) Rickettsioses are defined as infections with various species of Rickettsiae —bacteria-like micro-organisms, which are transmitted by lice, fleas, or ticks. With the exception of the epidemic, louse-borne typhus, which occurs only in man and is caused by *Rickettsia prowazeki,* all rickettsioses have an animal reservoir, from whence fleas or ticks can transfer it to man. Neither louse-borne typhus nor flea-borne typhus (murine typhus), which is caused by *R. mooseri,* has been reported in Kenya so far, although the natural reservoir of murine typhus is at least known.

Tick-borne rickettsial diseases appear to be relatively rare in Kenya at the present time. Systematic epidemiological investigations have not been carried out in Kenya since the first observations among Europeans in the Kenyan Highlands were reported by Gilks in 1920 (cit. from Craddok, 1974 a) who then had discovered cases also among the African population. But, at least in former times, there evidently existed some immunity, acquired during childhood, which prevented the occurrence of clinical manifestations at a later stage (Craddock, 1961).

Since an exact diagnosis can be made only after substantiation derived from serological investigations, hospital statistics under the prevailing conditions mean little—in so far as the diagnoses of suspected cases are recorded at all (see Fig. 7).

In East Africa rickettsioses are anthropozoonoses (Heisch, 1957, 1960), for ticks and their hosts constitute the natural reservoir of the rickettsiae. In East Africa

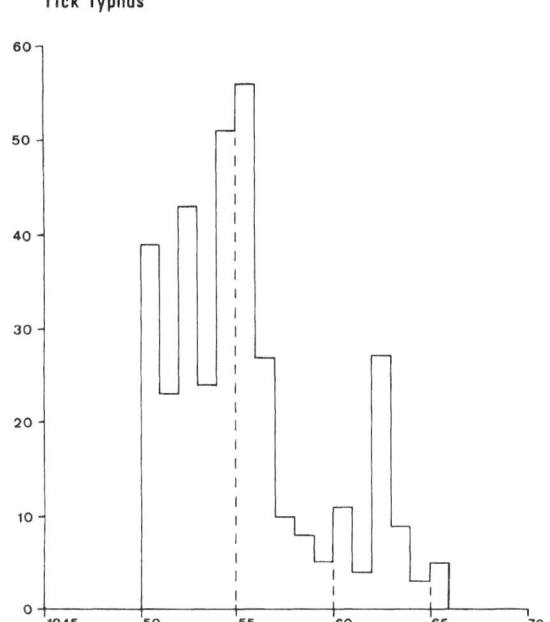

Figure 7. Tick typhus, number of annual notifications 1950 – 1972 (Source: Fendall and Grounds, 1965 c, Bulletin of Infectious Diseases, Ministry of Health, Nairobi, 1962 – 1972)

tick-borne typhus is caused by several variants of *Rickettsia rickettsii* (var. *piperi* and var. *conori*) ICD 0820.0, as well as *R. mooseri* (murine spotted fever) and *Coxiella burneti* (Q-fever) (Heisch et al., 1962). The rodent reservoir consists essentially of *Rattus rattus, Otomys angolensis, Arvicanthis abyssinicus, Lemniscomys striatus, Rhabdomys pumilio, Lophuromys aquilus* and *Aetomys kaiseri* (Heisch, 1961). The tick fauna is composed of *Haemophysalis leachi, Ripicephalus simus, R. sanguineus, R. pulchellus* and *Amblyomma variegatum* (Heisch, 1957, 1962). Details of regional variations in occurrence are not available.

Q-fever is caused by *Coxiella burneti*, another variety of rickettsia; since 1952 it has been occasionally diagnosed in Kenya. Evidently this anthropozoonosis is common among domestic animals, but without any major economic effect. The disease does not figure in veterinary medical reports. However, a special investigation did establish a more or less serious infection of dogs, camels, goats, and cattle. Human cases of Q-fever have been reported from Nairobi, Athi River, and the northern regions of Kenya (Craddock, 1974 b). More detailed studies by Vanek and Thimm (1976) have recently shown that Q-fever is present in man and cattle in all Kenya districts in which surveys have been made, in particular in the Eastern, Rift Valley, and Coast Province. However they assume that most of the adults, especially in the rural areas, are immune.

c) Relapsing fevers (Borrelioses transmitted by ticks and lice, ICD 087): Relapsing fevers are acute febrile infections caused by spirochaetes of the genus *Borrelia*, and are usually transmitted by ticks or body lice. They are characterised by febrile paroxysms lasting from one to

seven days, and followed by a febrile period of from two to fifteen days.

Similar to the rickettsioses, the tick-borne borrelioses were of some significance in Kenya, whereas relapsing fever transmitted by lice and caused by *Borrelia duttoni* probably only once occasioned a more serious outbreak — in and around Mombasa in the years 1945 – 46, in which case the infection had probably been imported on an Arabian dhow (Garnham et al., 1947).

The occurrence of tick-borne relapsing fever is essentially dependent on the way of life of the transmitters, which at the same time constitute an important reservoir. The vectors are made up of different species and sub-species of *Ornithodorus,* a leather tick of partly wild and partly domestic occurrence. The wild species of *O. apertus* live in the burrows of porcupines and groundhogs, *O. porcinus domesticus* and *O. moubata* in various domestic surroundings. *O. p. porcinus* and *O. p. domesticus* are the most common vectors in Kenya (Walton, 1962 cit. after Fendall, 1965 c). These tick species are highly dependent on the temperature and humidity of their habitat. The rapidly changing conditions of life and domestic hygiene have since the mid-fifties, contributed to the virtual disappearance of tick-borne relapsing fever.

In the cool highlands of Meru, Kikuyu, Taita, Digo, and Kisii, with precipitation in excess of 1,000 mm, tick-

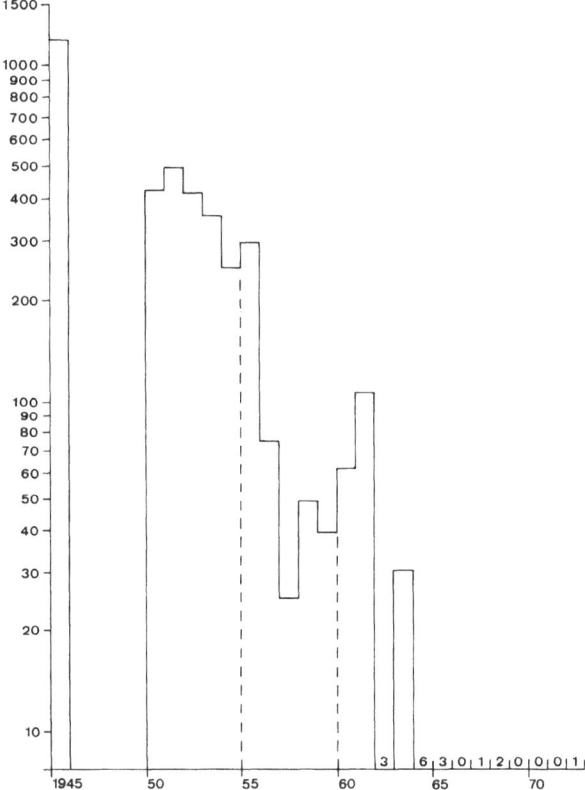

Figure 8. Relapsing fever, number of annual notifications 1945 – 1972 (Source: Fendall and Grounds, 1965 c, Bulletin of Infectious Diseases, Ministry of Health, Nairobi, 1962 – 1972)

borne relapsing fever posed a serious problem (Walton, 1955; Fendall, 1965 c). Thus children living in huts were particularly exposed, whereas adults, who spent more time outside, were less so. Cracks in the mud-built and urine-soaked sleeping quarters served as favourite hideouts for the ticks. They also particularly favoured the infection of infants, who had not built up any immunity (Walton, 1950; Bell, 1953).

A recession in the frequency of infections began with the forced resettlement of the Kikuyu following the Mau-Mau Rebellion in 1952–1955, when people were obliged to leave their tick-infested huts and to move into square, better apportioned and ventilated ones. The introduction of bedsteads, the application of DDT for the disinfection of huts, and a general and clearly visible improvement of residential and living conditions, all contributed to the virtual disappearance of the disease (see Fig. 8).

6. Filarial Infections Transmitted by Arthropods
(Filariasis, ICD 125)

Nematodes of the Filarioideae family are described as filariae; their females carry living larvae, so-called microfilariae, which are periodically or non-periodically found in the peripheral blood or skin, from which sites they are taken up by insect vectors during the act of blood-sucking. The adult filariae (macrofilariae) live as parasites, partly in the connective tissue and partly in the lymphatic system. The following species are differentiated: *Wuchereria bancrofti* (ICD 125.0) and *Brugia malayi* (ICD 125.1) which are the causal agents of elephantiasis, as well as *Loa loa* (ICD 125.2), *Mansonella ozzardi* (ICD 125.5), *Acanthocheilonema perstans* (ICD 125.4) and *A. streptocerca* (ICD 125.6), *Onchocerca volvulus* (ICD 125.3). With the exception of *Wuchereria bancrofti* and *Onchocerca volvulus*, the remaining filiariae are of subordinate importance in East Africa; that is, nothing is known about their frequency or distribution.

a) Wuchereria bancrofti (bancroftian filariasis): Elephantiasis *("matende")* and Hydrocele *("Mapumbu")* have been known in the Kenyan coastal zone at all times, although these symptoms had not been linked with one and the same cause. From this, and on the basis of his own extensive parasitological investigations, Wijers (1974) concludes that filariasis has been for a long time endemic in this region, whereas it does not appear to occur in the remaining parts of Kenya. Further endemic East African foci appear to be located in the Mwanza-Musoma region of Tanzania and the Lango and Teso districts of Uganda (Burkitt, 1951; Spencer, 1962), albeit without confirmation of this suspicion by precise epidemiological investigations as in the case of the coastal region.

Epidemiology: In Kenya Wijers (op. cit, 1974) investigated two epidemiologically distinct forms of *W. ban-*

crofti filariasis. The urban form is transmitted by *Culex fatigans,* which breeds in latrines and stings chiefly around midnight, and the rural form, which is transmitted by the *Anopheles gambiae,* the vector of malaria, and by *A. funestus,* which breed in reservoirs, pools, and other open waters and sting from before sunrise until sunset. In the Kenyan coastal area infections do not extend beyond an altitude of 400 m., and even in such areas they appear to follow river courses, leaving the level interfluves free.

Epidemiological investigations of filariasis involve certain inherent methodological problems, which render it difficult to make statements on the distribution of an infection among a population based on some isolated instance that has been discovered, such as a proof of microfilaraemia or proof of clinical or serological symptoms.

In 1962, the first comprehensive survey in Kenya was carried out by Nelson, Heisch and Furlong (1962) in the coastal region. Without specified selection for the random sample, 2.6% (Mombasa) – 26.3% (Malindi) of all persons over 15 years of age were found to be infested with microfilariae, with the male sex showing an infection rate about 10 – 15% higher than the female—as in all filariasis surveys. A connection between these conditions and clinical indications such as hydrocele among men or elephantiasis of the extremities among both sexes was not established (Tables 24 and 25). If proof of microfilariae in a population is regarded as indicative of endemic occurrence of *W. bancrofti,* without attaching a quantitative statement to this, it must be noted that the entire Kenyan coastal region constitutes an area of endemic *W. bancrofti.*

Over recent years more detailed investigations have been carried out, mainly by Wijers and his colleagues, which show that the degree of infestation of the coastal strip is significantly higher than had so far been assumed (Division of Vector-borne Diseases, Kenya Filaria Research Unit, Annual Report, Mombasa 1973). Systematic control measures, or mass prophylaxis or therapy, have not so far been applied, save for unspecific mosquito controls in tourist centres and towns. In this context it should, however, be noted that in the urban area *Culex fatigans* is becoming increasingly resistant to insecticides.

A sample survey, covering over 5,000 male over 14 years of age in Kwale and Kilifi District, recently published by Wijers, 1977, Wijers and Kinyanjui, 1977, and Wijers and Kiilu, 1977 revealed that the degree of infestation of the coastal region is significantly higher than had so far been assumed. Hereby microfilaria rate, microfilaria density, sign rate (i.e. rate of hydrocele or elephantiasis formation), resulting in a "filariasis index" corrected for absence of male in the sample were taken into account. Four main foci could be identified: a focus in the south, bordering Tanzania, one west of Mombasa, one just inland from Kilifi town and one focus along the Sabaki River, all of which had a filiarisis index of over 50%. A lower prevalence (below 50% filariasis index) was found in the coastal strip and in the hills north of Mom-

Table 24. W. Bancrofti Microfilaria Rates as Found by Examining Night Blood Films of Persons Aged 15 Years and Over on the Kenya Coast

Locality	Males		Females		Total	
	Number examinde	Percentage positive	Number examined	Percentage positive	Number examined	Percentage positive
Lamu area	272	27.9	367	20.9	639	23.9
Kipini and Witu	182	19.2	111	11.7	293	16.4
Tana river	223	16.1	213	13.1	436	14.7
Malindi and Mambrui	150	20.0	91	14.3	241	17.8
Malindi and Kakoneni	368	27.7	225	24.0	593	26.3
Kilifi area	247	16.2	187	9.1	434	13.1
Mombasa	305	4.2	235	0.4	540	2.6
Mombasa and Mariakani	334	18.2	149	12.1	483	16.3
Mombasa-Gazi	379	10.0	149	9.1	538	9.6
Msambweni-Vanga	299	11.0	80	3.7	379	9.1

Table 25. W. Bancrofti Microfilaria Rates in Relation to Elephantiasis and Hydroceles in Persons Aged 15 Years and Over on the Kenya Coast

Locality	Sex	Number examined	Percentage W. bancrofti	Percentage elephanti-asis leg	Percentage hydrocele
Faza	male	89	40.6	16.8	39.3
	female	132	31.1	6.7	—
Ganda and Kakoneni	male	136	28.3	5.8	34.5
	female	104	31.3	1.9	—
Jomvu and Mariakani	male	107	14.4	0.6	19.6
	female	156	14.5	1.3	—
Vanga	male	109	31.6	6.4	21.1
	female	102	33.3	1.9	—

(Tables 24 and 25 from Nelson, G. S., Heisch, R. B. and Furlong, M., *Trans. R. Soc. trop. Med. Hyg.,* 56, 202 – 217, 1962).

basa, i.e. in the areas with the highest rainfall and the highest population density. A still lower filariasis index was found in the sparsely populated area of Lamu District and along the Tana River, although the elephantiasis rate was found to be comparatively high.

b) Onchocerciasis (ICD 125.3). Onchocerciasis was formerly endemic in six well-defined areas of western Kenya. The familiar symptomology of itching skin and "river blindness" made it clear that this disease had occurred for a long period. Systematic investigation of the disease was begun in 1921 and followed by what has remained, until now, its successful eradication. Onchocerciasis in Kenya is one of the few examples of successful extermination of a medically-significant insect vector over a long period.

In Kenya the only vector was *Simulium (lewisellum) neavei.* This species of Simulium was discovered by S. A. Neave in 1911 on the Yala River in the Busia District. Its significance in the transmission of onchocerciasis became evident only twenty years later. Systematic investigations led to the discovery of foci of onchocerciasis in Kenya, situated in Kaimosi, Kakamega, Yala in Kakamega District, in Kodera in Kericho District, and in Riana and Ngoina in Kisii District. *S. l. naevei* was the vector in all these areas, totaling more than 8,000 square kilometres (see Table 26) (Buckley, 1949, 1950).

Extensive investigations along the southern and eastern slopes of Mt. Elgon, in the Cherangani Hills, and on the Mau Escarpment failed to provide evidence of the occurrence of *S. neavei* or the anthropophilic simulium flies. Nonetheless, other species of simulium do exist. Investigations in the area of Mt. Kenya, the Aberdares, and the Niambeni Range also failed to provide proof of *Simulium neavei.* The western slopes of Mt. Elgon near Mbale and one area on the southern slopes at Malakisi on the Kenyan border do, however, still have massive infestation by *S. neavei,* and onchocerciasis is equally common there.

There is a twin species of the *Simulium damnosum* complex in Kenya, which is still morphologically indistinguishable from the anthropophile one, although it evidently takes no part in the transmission of onchocerciasis (Highton, 1974 c). The occurrence of *S. (L.) neavei* in Kenya was restricted to shaded, fast-flowing perennial streams and rivers in hilly and mountainous terrain. The breeding places are dependent on the presence of the fresh-water crab *(Potamonautes niloticus)* (M. Edwards), which prefers sunny places in the water. The larvae of the simulium live in the shells of the crabs in well-aerated water. The adult simulium has a relatively short flying range, so that it always remains within the area offering conditions favourable to its life (Buckley).

The first control programmes were carried out by Garnham and McMahon in the Konder area in 1946 by

Table 26. Distribution of Onchocerciasis

District Locality	Area	vector	Population infection rate	Eye complication rate	Infection rate of mosquitoes	Control program carried out	No fresh infections result among children later born (1966)
Kenya:							
Kakamega							
Kaimosi	5,000 km²	S. neavei	72%	10%	10%	1947/48	,,
Kakamega						1954/56	,,
Yala							
Kericho							
Kodera	170 km²	S. neavei	49%	10%	10%	1946	,,
Kisii							
Riana	42 km²	S. neavei	21%	1.6%	10%	1947	,,
Ngoina	3,000 km²	S. neavei	29%	1.8%	10%	1952/53	,,

Source: McMahon et al. (1958); Brown (1962).

the application of a DDT emulsion. Over the period 1952 – 54, McMahon arranged for this treatment to be applied to all streams and rivers within an area of 8,500 square kilometres, inhabited by more than a quarter of a million people (McMahon et al., 1958, Table 26). Investigations carried out by Nelson and Grounds in 1958 and by Robert et al. in 1967 showed that no new infections had occurred in the age groups born after the elimination of the vector. However, a reservoir of microfilariae for anthropophile simulium flies, which still possibly survive, remained among the older population. Even now this campaign may still be the most effective one ever to have been carried out for the long-term control of insect-borne diseases.

S. naevei still exists in some areas, especially in the focus near Malakisi, which is linked with the uncontrolled focus on the western slopes of Mt. Elgon. Highton, (op. cit., 1974) has given an urgent warning of the danger of the re-introduction of onchocerciasis as a consequence of systematic irrigation measures and hydroelectric power station projects.

II. Infectious Diseases Usually Transmitted Directly from Man to Man (Contact and Air-borne Diseases)

1. Tuberculosis
(ICD 010)

Definition: "A usually chronic infectious disease caused by *mycobacterium tuberculosis* or, more rarely, *M. bovis,* usually acquired by the inhalation of infectious material and sometimes by ingestion, especially in the case of *M. bovis.* It attacks most commonly the lungs and may spread by haematogenous dissemination to other parts of the body. The bovine form generally affects extrapulmonary organs (intestines, bones, lymph nodes, etc.)". *

* C.I.O.M.S. Communicable Diseases, Provisional International Nomenclature S. Btesh. (Ed.) Geneva, 1973

Tuberculosis is one of the greatest health problems in Kenya, as in all developing countries. The first epidemiological investigation in the field was conducted in the year 1948/49 (Haynes, 1951) during which almost 50,000 tuberculin tests were carried out in 18 localities in Kenya. The second investigation took place ten years in 1958/59 later and was organised by the Ministry of Health and supported by the W. H. O. (Roelsgaard and Nyboe, 1961); on this occasion, a representative sample of 8,500 persons (0.7% sample) was examined on the basis of the population distribution of the census in the year 1957. This recent investigation did not permit unambiguous estimates of regional differences in distribution to be made, since these were only given at the provincial level. It was pointed out that the volume and selection of the samples did not permit reliable regional differentiation (see Fig. 17, back of Map 9)—this had not been the aim of the investigation—but rather the verification of tuberculosis infection in the different age groups. It was found that 3% of the children in the age group below 5 years already had had some contact with tuberculosis and probably still had active tuberculosis at this age. About 13% of the children in the age group 5 – 9 years and more than 20% of the age group 10 – 14 years showed a positive Mantoux reaction. In the age group above 20 years, the male population was predominant, a fact attributable to the greater risk of infection at work outside the family. More than 3% of the age group over 10 years showed changes in their lungs following x-ray detection and were therefore suspected of having tuberculosis.

On the basis of the results it was estimated that in 1963 about 110,000 inhabitants of the 6 million were suffering from tuberculosis, 40% of whom in turn presented a source of infection; out of these about 38,000 were children below the age of 5 years. In the age group above 10 years 0,6% were found to be bacteriologically positive in terms of smears and cultures (Table 27).

Furthermore, the investigation showed the greater risk of infections among children below 5 years of age in those households with infected family members (26%

Table 27. Tuberculosis Survey in Kenya (after Roelsgaard and Nyboe, 1961)

Provinces		Percentage of infected children aged 0–14 years	Percentage of persons with tbc lung pathology	reported 1957/59 per 1,000
Rift Valley		11.5	6.77	0.74
Nyanza		7.5	9.00 *	0.45
Central		13.0	8.50	1.97
Southern		15.1	8.22	0.85
Coast	Kwale Kilifi	9.9	3.85	1.42
Total		10.7	7.88	1.10

* Central Nyanza 14.96
  Northern Nyanza 7.67
  Southern Nyanza 7.93

Table 28. BCG-Vaccination in Kenya 1962, 1966, 1970, and 1973 *

	Total vacc. 0–16 per year	cumulative No. of vacc. 0–16	cumulative coverage 0–16 (%)
1962	20,961	20,961	0.5
1966	723,607	1,868,640	33.5
1970	729,970	4,447,826	61.4
1973	572,608	6,644,537	79.1

* For details see Fig. 9.

positive Mantoux reactions), as compared to 2.4% among children in households without infectious members of the family. In the age group 5 – 9 years the frequency of positive Mantoux reactions of both groups has already risen to 70%, and 12% respectively. The difference is less among older children because of the increasing additional exposure to tuberculosis outside the family.

Although the reported figures, whether furnished by the Annual Return of Diseases from the government hospitals or by the Notification of Infectious Diseases from the district medical officers, depend among other things on the availability of diagnostic facilities and above all the awareness of tuberculosis, it is possible to discern characteristic regional differences. Geomedical analysis of the hospital recording rate for tuberculosis in 50 hospitals within central and western Kenya has shown that the frequency of tuberculosis in hospitals located in cool and dry climatic regions was ten times higher than in hospitals in humid-warm or dry-warm climatic regions (Diesfeld, 1969 a).

The Tuberculosis Control Programme: On the basis of the Ministry of Health's tuberculosis survey carried out with assistance of W.H.O. in 1958 and 1959, an action programme was devised which primarily provided outpatient treatment using thiacetazone/isoniacid and inpatient treatment using streptomycin/PAS/pyracinamid for seriously infected patients. A follow-up of this programme showed a cure rate of 63%, although a primary

isoniacid resistance of at least 5.2% must be assumed in Kenya (Kent, 1974).

In the early sixties, the first BCG vaccination programme was inaugurated by the Ministry of Health and supported by the W.H.O. and U.N.I.C.E.F., in the course of which 3.7 million children below the age of 16 were vaccinated during the 8 year period from 1962 – 69. Since 1970, this campaign has been combined with a National Smallpox Eradication Programme. By the end of 1973 a further 2.9 million doses of BCG vaccine had been administered (Fig. 9 and Table 28).

The programme is continuing on the basis of 22 regional sectors and under the direction of the Communicable Disease Control Division of the Ministry of Health. It is assumed that 500,000 – 600,000 annual revaccinations will effactually make up for the reduction in the collective protection of vaccination caused by the later-born population. Since an estimated 15% of newborn babies are seen in ante-natal clinics, 25% of children aged between 1 and 5 in child welfare clinics and nursery schools, and 70% either in school examinations or OPD-attendance at dispensary health centres and hospitals,

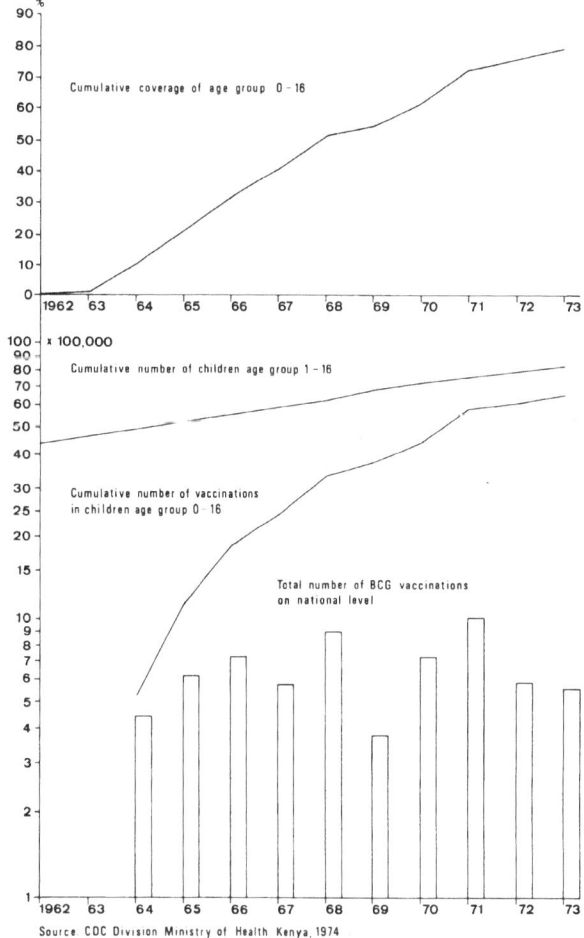

Figure 9. Tuberculosis vaccinations, 1962 – 1973.

where regular vaccination sessions are conducted, it may be assumed that an 80% coverage can be maintained.

It can be assumed that every child born in the coming years will receive BCG vaccination at one of these institutions in the course of time. Zonal supervision and permanent random assessments are carried out by independent teams. At the same time the treatment programmes continue in accordance with the accepted standards. If it is possible to maintain this national programme as planned, it can be assumed that within a generation the problem of tuberculosis will be under control in Kenya.

2. Leprosy
(ICD 030)

Definition: "chronic infective disease caused by *Mycobacterium leprae,* and characterized by a long incubation period and slow onset with vague pains and areas of paraethesia followed by typical lesions on the skin, mucous membranes, and nerves. Two polar forms are described, namely lepromatous and tuberculoid leprosy; intermediate, borderline and indeterminate forms are also recognised."

Approximately 1% of the East African population suffers from leprosy (Schaller, 1970). At present there is no conclusive evidence which can explain the considerable regional differences of occurrence, which undoubtedly exist. The first large-scale investigations of the occurrence and distribution of leprosy in Kenya were carried out by Innes (1949); the methodological short-

Table 29. Frequency of Leprosy — reporting from "Weekly Notifications of Infectious Diseases" (median annual notifications for 1963 – 1972 per 100,000 1969-Census-population) in descending order

District/Province	Leprosy Notifications per 100,000 population	
Area of high endemicity:		
1. Busia District	95.50	
2. Kisumu/Siaya District	73.05	I
3. Kwale District	65.36	$\geqq 50/100,000$
4. Kilifi District	23.05	II
5. Bungoma District	17.68	10–49/100,000
Area of medium endemicity:		
6. Kakamega District	6.01	
7. Lamu District	4.54	
8. Kitui District	3.79	
9. Taita-Taveta District	2.70	III
10. Meru District	2.54	1–9/100,000
11. Embu District	2.23	
12. Central Province	1.55	
13. Mombasa	1.21	
Area of low endemicity:		
14. South Nyanza District	0.58	
15. Rift Valley Province	0.49	
16. Machakos District	0.42	
17. Kisii	0.20	IV
18. NE Province	0	<1/100,000
19. Tana River District	0	
20. Kisii District	0	

comings contained in that work scarcely allow an interpretation to be made, although they do draw attention to crucial points in western Kenya and on the coast, which were confirmed and more accurately investigated in western Kenya through the more exact sampling methods of J. A. K. Brown (1959) between 1951 and 1959.

Fendall and Grounds (1965 b) estimate that three-quarters of all the 35,000 cases of leprosy assumed in Kenya — other estimates put the figure at 60,000 – 70,000 (Verhagen, 1974) — occur in this region.

However, systematic investigations carried out in other parts of Kenya between 1967 and 1968 show new, and so far unknown, foci in Kitui District, in the south of Embu District, and in eastern Meru District. A revision of the known leprosy areas in the Western Province using the same sampling methods resulted in a prevalence of 1.2 to 1.4% in two locations in the Kakamega District, of 3.2% in Busia, and of 1.4% in South Nyanza (Ziedses des Plantes et al., 1968).

Although the reporting is definitely incomplete, the regional distribution of the frequency of leprosy can, at least its tendency, be estimated from the Weekly Notifications of Infectious Diseases published by the Ministry of Health. From the median number of cases for the period 1962 – 1972 related to the census population of 1969, foci can be recognised which have been confirmed by various special leprosy investigations (see Table 29).

The median number of hospital leprosy cases reported in the "Annual Return of Diseases" over the period 1963 – 1972 and related to the estimated population in the catchment area (Census 1969) portrays essentially the same tendency, even if some hospitals in northern Kenya mention more leprosy than appears in the Weekly Notifications. On the other hand, no cases of leprosy are registered by the hospitals at Kwale and Msambweni, although it is well known that leprosy occurs particularly frequently in this part of the Coastal Province (Ministry of Health, Annual Report Coast Province, 1973).

Three clearly discernible grades of endemicity can be distinguished on the basis of information at present available, with a binding statement concerning prevalence only from individual random tests (see Table 29). So far as particular details are concerned the various leprosy foci can be characterised as follows:

a) Foci in western Kenya: Different investigations agree in locating the highest leprosy infection rates in the country (1 – 3%) in the Kisumu, Busia, and Bungoma Districts. One of the two large hospitals for leprosy patients in Kenya is to be found in Alupe near Busia; contrary to the "Weekly Notifications" and the "Annual Hospital Returns" this focus in the area of the Lake Victoria littoral extends further to south Nyanza (Leiker, 1966; Ziedses, 1968), thus wholly embracing the settlement area of the Luo, a Nilotic tribe which immigrated from the north (Kisimu 93%, Siaya 96%, South Nyanza 89%, according to the 1969 Census). In Busia District the proportion of Luo is smaller, but 31% belong to the equally Nilotic Iteso tribe. In Bungoma District the Nilotic Iteso and Sabaot compose 15% of the population.

Towards the north, in Bungoma District, and towards the east, in Kakamega District, the frequency of leprosy appears to decrease. In Bungoma, Busia, and Kakamega the Luhya, members of the Bantu group, predominate. In his investigation of the Ugandan and Kenyan "leprosy belt", Brown (1959) suspects the existence of a genetically determined factor promoted by in-breeding which predisposes the geographically and ethnically isolated Nilotics as well as the displayed Luhya (affected by the inmigrating Nilotics) to leprosy, thus perhaps explaining this focal concentration of leprosy. The precise delimitation of leprosy towards the east, an area in which leprosy is evidently almost non-existent, is indeed remarkable.

b) Foci in the Coastal area: In the coastal area recent investigations (Hartmann, 1973) have demonstrated a rather variable frequency. Whereas leprosy does not appear to be a special problem in Lamu District and on the Tana River, in Kilifi and Kwale Districts an infection rate of about 1% of clinically diagnosable leprosy is usual. The differentiated investigations mentioned above show that the Waduruma and other Mijikenda groups as well as the Wakamba, who have moved into this region, and those groups in more isolated areas appear to be particularly affected. Among the non-Bantu groups in the northern coastal area leprosy seems to occur less frequently, indeed to be retreating.

c) Foci in the Taveta area: Among the *Taveta,* too, leprosy appears to be more frequent than is the case with the surrounding Taita (Innes, 1949), a fact clearly reflected in the hospital recording rates of Taveta (59/10,000) as compared to those at Wesu in the Taita Hills (2/10,000). There is no obvious explanation for the rather striking differences.

d) Focal areas in the Central Lowlands of Kenya: Leprosy foci that had long escaped notice were found in the Kitui District, in the south east of Embu District, and in the east of Meru District (Tharaka, Nyambene); these have a frequency rate in excess of 1% (Ziedses et al., 1968). This observation is also reflected in the Notifications and Annual Returns, which stand out clearly against the neighbouring regions.

In the Central Highlands (Central Province) leprosy appears to occur relatively rarely. The various districts of the Rift Valley Province, together with the northern and north eastern districts, report very low figures or nil returns. Nevertheless, more cases of leprosy are registered in the Annual Returns of Diseases from the Lodwar, Marsabit, and Isiolo hospitals than in the District Notifications of Infectious Diseases. It may at least be stated that leprosy evidently does not play as special a role in these areas, as it did in the four foci of leprosy described above.

e) Prevention and eradication: After the Second World War leprosy settlements gradually disappeared. In 1948

Innes introduced chemotherapy. Apart from Alupe, there is only one other hospital for leprosy patients remaining at the present time, located at Tumbe near Msambweni on the Indian Ocean coast. Alupe also houses a leprosy research station under the aegis of the East African Common Services Organization. After 1962 the treatment of leprosy in the form of open treatment was transferred to government hospitals, albeit with disappointing results. It remains to be seen whether the national campaign against tuberculosis will be able to control leprosy with BCG vaccine, which is thought to provide immunization against leprosy as well.

3. Smallpox
(ICD 050)

Smallpox is an acute infectious disease caused by the smallpox virus *variola major* which, owing to its highly contagious nature, gives rise to continuously recurring epidemics among unvaccinated populations.

At the initial stage, transmission results from droplet infection, at the pustular stage, by smear infection from person to person. The virus can survive for long periods in dust.

Until systematic immunization of the population was carried out in 1972, smallpox was endemic in Kenya, although the number of smallpox cases had declined over the years and despite the continuous improvement in the quality of smallpox registration and the thereby increased number of official registration of cases (Fig. 10).

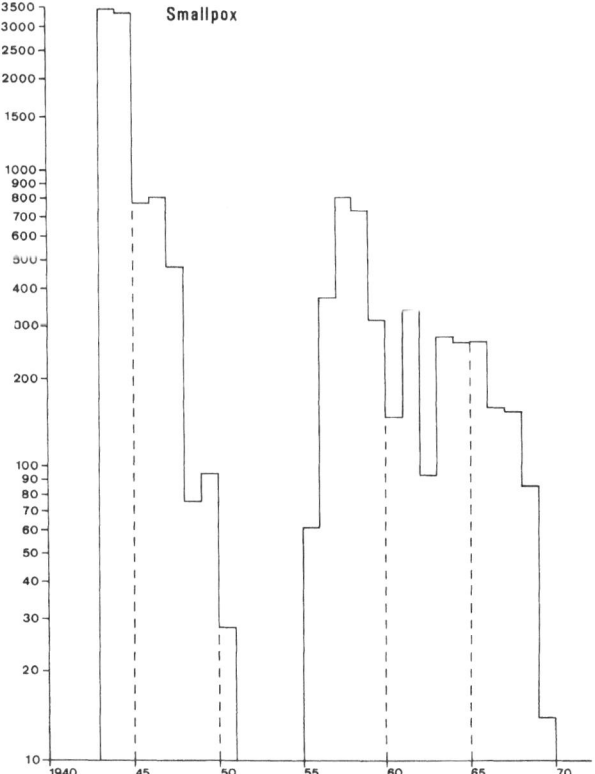

Figure 10. Smallpox, number of annual notifications 1942–1972 (Source: W. Koinange Karuga, 1974 and Fendall and Grounds, 1965 b)

Smallpox assumed epidemic proportions during and after both the world wars, during the recession of the thirties, and during the Mau Mau Revolution. In Kenya smallpox had a fatality rate of 1 – 2%, indicative of a relatively non-virulent virus in the main, infection by *variola minor*. Clinical cases occurred almost solely among non-vaccinated persons. Fairly extensive vaccination campaigns and voluntary vaccinations against smallpox were not enough to bring the smouldering endemic state to an end. Small, localized outbreaks effected an immediate improvement in people's readiness to be vaccinated, but this declined as the outbreak subsided.

The distribution of officially registered, and so far as hospitals are concerned, reported cases agrees in showing lower frequency in areas of comprehensive vaccination, covered in the reports of the district medical officers.

Seymour-Price et al. (1960) regarded the variable accessibility to the transport network as an important reason for the variation in the vaccination rate, and, correspondingly the variation in the smallpox frequency. Fendall and Grounds (1965 b) noted, on the other hand, that smallpox occurred more frequently in the settlements of plantation workers and other population groups living close together, although vaccinations could have been carried out quite easily in such places without undue communications difficulties. Nevertheless, smallpox has occurred less frequently in Kenya over recent years than was the case in neighbouring countries. The last cases of imported smallpox in Kenya mainly entered via the nothern borders, whereas over the past twenty years the coast and airport have ceased to provide points of entry.

In the case of reported, suspected or actual smallpox, a mobile unit of the Ministry of Health is now available to investigate the outbreak immediately and to take suitable isolation and vaccination measures. After appropriate planning, a national smallpox vaccination programme was carried out in 1972, in the course of which 13 million vaccinations and re-vaccinations were administered. Nine million of these were carried out by mobile vaccination teams, four million by static units. The programme was combined with a BCG vaccination programme. An independently conducted investigation of the outcome of vaccinations in five provinces showed that at the campaign's closing date an immunization rate in excess of 80% of the population had been achieved, which in the age group 5 – 14 years had, in some areas, been increased to more than 90%. In order to maintain the success of vaccinations it was followed up by a "maintenance phase" in June, 1972, during which one mobile unit for each of the 22 sectors of the country was to repeat the performance within a year (Koinange Karuga, 1974). It can therefore be assumed that under these conditions Kenya will remain free from smallpox infection, even if cases of the disease are imported from neighbouring countries.

Figure 11. Meningococcal meningitis, number of annual notifications 1950 – 1972 (Source: Fendall and Grounds, 1965 b, Bulletin of Infectious Diseases, Ministry of Health, Nairobi, 1962 – 1972)

4. Cerebrospinal Meningitis
(ICD 036)

Cerebrospinal meningitis (CSM) (meningococcal meningitis) is an acute purulent inflammation of the meninges due to *Neisseria meningitidis*, often occurring epidemically in certain tropical areas and sporadically with small outbreaks all over the world.

Extensive epidemics of meningitis, such as occur in the area of the "meningitis belt south of the Sahara" (Jusatz, 1961), have hitherto not been observed in Kenya. Smaller outbreaks, which affect few people, do occur every year (Fig. 17, back of Map 9), although they do not manifest the essential seasonal peaks, and those affected are mainly children (Fendall and Grounds, 1965 b). As yet epidemiological investigations in more detailed form have not been carried out. *Neisseria meningitidis* plays a minor role in bacteriological specimen investigated at the National Public Health Laboratory, Nairobi, which are sent in from Kenyatta National Hospital and from surrounding hospitals (Say, Itotia and Wamola, 1974).

The Weekly Notifications of Infectious Diseases" for 35 districts, covering the period 1963 – 72 and relating to 100,000 inhabitants of the median year 1967, show that

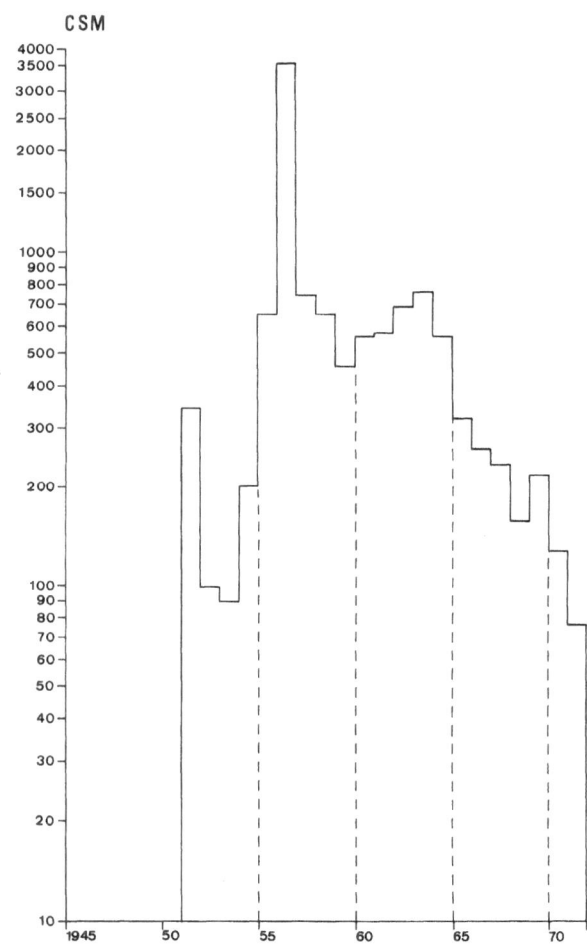

the Central Province has the highest annual rate of registrations with 6.69 cases per 100,000 inhabitants; it is followed by the Eastern Rift Valley and Western provinces, each with less than half the number of reported cases. The Coast Province, Nyanza Province, and Nairobi each report about one-tenth of that number of cases. Although these figures cannot be taken as absolute values, they do show the varying distribution, which possibly shows a connection with the higher altitudes of the settlements, in which relatively long sojourns within the dwellings, due to climate, and high settlement densities produce closer contact and with it increased risk of transmission (see Fig. 17, back of Map 9 and Fig. 11).

5. Yaws
(ICD 036)

Yaws is a chronic treponematosis, caused by *Treponema pertenue* which is transmitted by direct contact and, much as in the case of cosmopolitan syphilis *(T. pallidum)*, leads to primary lesions of the skin as well as to secondary and tertiary destruction of skin and, above all, bones. In Kenya, as in the rest of Africa, it was once very common. The Annual Reports of the Medical Department reported on it from the twenties onward, with the numbers quoted there increasing at the same rate as treatment possibilities became available. In the period 1921 – 1932, tens of thousands of cases were treated every year, at first with Neosalvarsan and later with bismuth preparations; since the early fifties treatment has been with long-term penicillin (PAM). During the period when penicillin was being introduced, the numbers of those seeking treatment again showed a considerable rise, but since the midfifties their numbers have progressively declined. The reputation which Western medicine already enjoyed at an early stage can in essence be atributed to these successes (Fendall and Grounds, 1965 b; Vogel and Huma, 1974).

It was, however, very soon recognized (Annual Report, Medical Department, Nairobi, 1931) that even minimal improvements in personal and general conditions of hygiene—that is cleanliness—are no less important than chemotherapy in the control of this disease. Through these means, as well as through the specific and general distribution of penicillin through official and even unofficial channels, yaws is to all intents extinct in Kenya today. When checking the figures on yaws reported in the "Annual Return of Diseases" of the hospitals in the field, discussions with physicians showed that the registered cases appear to be relics of tertiary yaws, and that no new cases had come to their notice for years. It is only among the nomadic herdsmen in northern Kenya that some isolated fresh cases may still be observed (Vanek, 1968, personal communication); as regards hygienic conditions in particular, the herdsmen are to be considered a marginal group.

The assumption that there is a cross-immunity between yaws and veneral treponematosis (syphilis) is contradicted by the fact that the disappearance of yaws has rendered the population more receptive to syphilis (Vogel and Huma, 1974).

6. Trachoma and Other Infections Causing Blindness
(ICD 076)

Scarcely any other disease is as favoured by lack of water and personal hygiene, by ignorance and poverty as trachoma, an infectious keratoconjuntivitis, caused by *Chlamydia trachomatis,* transmitted by contagious infection under poor hygiene conditions, especially in hot and arid areas of Africa and Asia.

Due to inadequate specialist diagnostic conditions, trachoma and its consequences cannot always be clearly distinguished from other chronic eye infections and their aftermaths. It is therefore wise to treat pronouncements on trachoma distribution with a certain scepticism. Medical statistics undoubtedly present a considerable number of eye infections under the heading of trachoma, which are in fact of bacterial or viral origin.

As soon as there is an improvement in conditions of personal hygiene and the possibility of washing the face each day with clean water becomes a reality, the frequency of chronic eye infections promptly decreases. Eye infections are thus closely connected with the availability and usage of water for washing, and are thus not only dependent upon hydro-geographical and climatic factors, but above all on socio-economic and socio-cultural factors. Investigations carried out in Kenya have provided a clear illustration of this situation (Bisley and Burkitt, 1974).

In Kenya, infections of the eye are most widely spread among children, and women are more frequently affected than men. Bisley and Burkitt (op. cit.) suspect that boys escape from the unhygienic domestic milieu at an earlier stage, and once outside it are more likely than girls to have the chance to take a bath. For example, in West Pokot the authors found eye infections among 98% of the population, whereas schoolchildren in the same area at a school near a river only manifested a 30% infection rate. Stones (1939) and Henderson (1968), both in Uganda, estimated that about 15% of the children are infected by trachoma, especially in semi-arid areas; Emiru and Dechet (1969), observed a frequency of 18% among the population in Uganda examined by them, and these same authors attributed 40% of the blindness to trachoma. According to Calcott (1954) 6.3% of 16,000 schoolchildren in Kenya were found to be suffering from trachoma when examined between 1953 and 1955. Trachoma is widespread in Kenya, but it appears to take a relatively mild course so long as it is not aggravated by additional infections. This may explain the fact that only 3.7% of blindness is unequivocally caused by trachoma (see Table 30).

The Opthalmological Survey of the British Empire Society for the Blind and of the Ministry of Health,

Table 30. Causes of Blindness among 1,093 Blind Persons in Kenya, 1953 – 1956 (after Calcott, 1954).

	Cases	%
Senile cataract	477	43.6
Panophthalmitis	183	16.7
Ulceration of the cornea	182	16.6
Optic atrophy	62	5.7
Glaucoma	50	4.6
Trachomatous scarring	40	3.7
Non-senile cataract	25	2.3
Irido-cyclitis	23	2.1
Injury	17	1.6
Onchocerciasis	15	1.4
Retinopathies	13	1.2
Degenerative conditions	6	0.5
	1,093	100

Kenya, conducted during the period 1953 – 1958 (Calcott, 1954 op. cit.), in which 7,747 persons were examined, produced 3,248 (=42%) normal and 4,499 (58%) pathological findings. Of these, 1,677 (i.e. 37% of the pathological total or 21% of the total number of persons examined) had to be attributed to trachoma. Calculations have shown the rate of loss of sight to be 12/1,000 of the population or, relating this to the then total population, the probability of 70,000 blind persons. The main causes were senile cataract (43.6%) and the aftermath of infections (45%) (see Table 30).

This report demonstrates that 80% of the causes of blindness can be prevented and consequently the decisive measures for its control are in the sphere of health education.

The regional distribution of the median annual illnesses classified as trachoma in the Annual Return of Diseases reported by 54 Kenya Government hospitals over the past ten years and calculated as a hospital recording rate for 1963 – 1972 provides (in accordance with the ideas advanced above) more of an indication of the conditions favouring trachoma and eye infections than a picture of trachoma distribution. In this semi-quantitative form, the presentation of this geographical frequency distribution takes on an impressive image. According to it, the greatest frequency occurs in the northerly areas with nomadic herding population. 10 of the 54 hospitals that were investigated (i.e., 18%) are situated north of a line linking Kapenguria—Tambach—Isiolo—Meru—Wajir. The hospitals next in order of precedence (30%) are located exclusively in central Kenya, together with the southern Rift Valley and Wesu (see Table 31).

The hospitals in the coastal area and western Kenya, Nyanza, the Western Province, and the central Rift Valley represent the lower half, with small to non-existent numbers of trachoma cases reported. This unambiguous distribution permits one to draw conclusions concerning conditions of hygiene and other factors favouring or restricting trachoma.

III. Infectious Diseases Usually Transmitted Indirectly from Man to Man by Water and Food (Water- and Food-borne Diseases) and Tetanus

In this chapter the most important water-borne and food-borne diseases are combined, although their actual distribution and significance are relatively little known. From an epidemiological viewpoint, this group includes —apart from the classical bacterial and protozoal enteritic infections such as cholera, typhoid fever, enteritic fever, and other salmonella, shigella and amoebic dysenteries—virus diseases such as poliomyelitis and virus hepatitis. Intestinal anthrax, although it could be included on etiological grounds, will be dealt with in the section on anthropozoonoses. Because of its close association with the conditions of soil, tetanus will be dealt with in this section.

1. Poliomyelitis
(ICD 045)

Generally not much is known about the distribution of poliomyelitis, since no epidemic outbreaks of the para-

Table 31. Hospital Recording Rate (HRR) distribution for trachoma

Number of reported cases of trachoma	Number of hospitals	HRR per 100,000	Hospitals
very frequent rank numbers 7, 8, 9	10 (=18,5%)	200 – 250	Isiolo, Marsabit, Maralal, Meru, Moyale, Kapenguria, Lodwar, Wajir, Mandera, Molo
frequent rank numbers 5, 6	17 (=31,5%)	30 – 170	Kiambu, Kerugoya, Thika, Tambach, Machakos, Embu, Narok, Wesu, Nanyuki, Kajiado, Lokitaung, Londiani, Murang'a, Kabarnet, Nyeri, Kitui, Kapsabet
infrequent rank numbers 1, 2, 3, 4	27 (=50%)	0 – 29	Mombasa, Nakuru, Thomsons Falls, Nairobi, Taveta, Voi, Garissa, Lamu, Kipini, Naivasha, Nandi, Kitale, Kapkatet, Eldoret, Bungoma, Kisii, Malindi, Kisumu, Kilifi, Galola, Kwale, Msambweni, Makindu, Tigoni, Kericho, Kakamega
	54 (=100%)		

lytic form occur. Mainly spread by faecal-oral transmission, this entero-virus infection attacks children in those places having generally low conditions of hygiene and where the majority of such infections escape detection and paralytic cases occur relatively seldom. From an epidemiological point of view this state of high endemicity leads to a good state of immunity in later life. In the tropics antibodies for all three types of polio virus are already found by the age of six in most cases. As the standard of living rises, the exposure to infection at an early age becomes less and paralytic forms of the infection will become more frequent, thereby leading to the classical form of "epidemic poliomyelitis".

Kenya has entered this epidemiological situation in the mid-1950's approximately. Whereas paralytic cases of poliomyelitis had scarcely ever occurred beforehand, an increasingly well-developed 3-year cycle of epidemic poliomyelitis set in from 1954 onwards (see Fig. 12). As a rule the epidemics occurred in September and October. In this manner epidemic peaks were observed in 1954, 1957, and 1960 (Fendall and Lake, 1958, Fendall, 1962; Fendall and Grounds, 1965 b; Kaur and Metselaar, 1967).

The 1960 epidemic clearly demonstrated regional differences (see Fig. 17, back of Map 9), in which Nairobi, the Central Province, and the Rift Valley Province produced the highest number of reported cases, whereas the Western, Northern, and Southern Provinces were less affected (see Fig. 17, back of Map 9). Even if this distribution is indicative of varying medical services and although the Western Province is comparatively better supplied than the Rift Valley Province, it is certainly —based on figures reported by hospitals—an expression of varying standards of development and hygiene within Kenya.

Virological investigations carried out in the period 1965 – 1969 have shown that of 559 poliomyelitis patients whose age was known and in whom polio virus could be isolated, 99% were under the age of six. No patient was older than 20 years.

Of 230 serum samples, which had been collected at different locations in Kenya over the period 1965 – 69 from children who had not had poliomyelitis, more than 80% of the 4 to 5 year olds and more than 90% of the 10 to 11 year olds had already developed antibodies against all three types of polio. On the other hand, among a total of 868 polio patients type 1 was isolated in 84% of all cases, type 2 in 10%, and type 3 in 6% of all cases (Nottay and Metselaar, 1973).

The epidemic predicted for 1963 on the basis of previous experience was prevented by the implementation of a massive campaign of oral vaccination against polio in the spring (Fendall and Grounds, 1965 b). On the basis of the virological and epidemiological investigations of previous years apparently successful methods were developed in campaigns of specific mass vaccinations under the prevailing communications facilities, public health services, finance, and staffing problems (Metselaar et al., 1973).

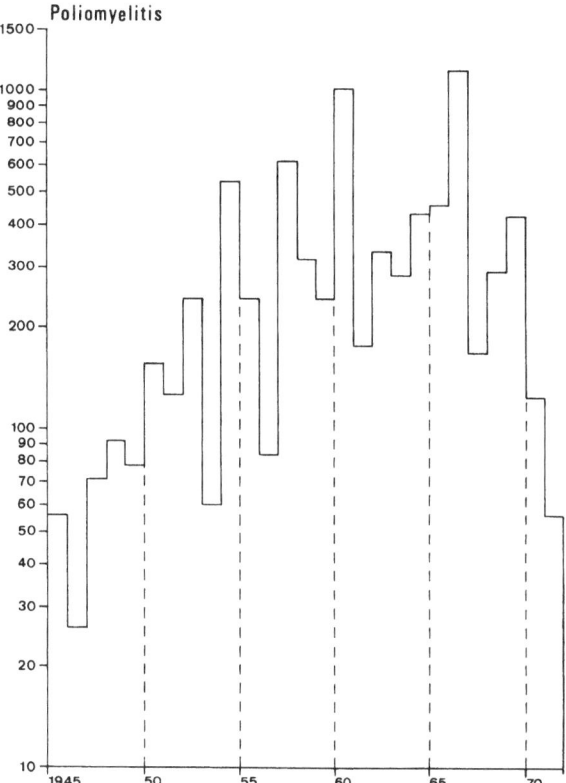

Figure 12. Poliomyelitis, number of annual notifications 1945 – 1972 (Source: Fendall and Grounds, 1965 b, Bulletin of Infectious Diseases, Ministry of Health, Nairobi, 1962 – 1972)

Since epidemiological surveillance suggested that a new epidemic was to be expected in spring, 1972, a pilot project for a national vaccination campaign was carried out on the basis of these earlier experiences. In this campaign it was demonstrated that a vaccination programme could be carried out in every province within two days with the support of lay assistants and all the public services and that such an interval method is to be preferred to long-term, continuous programmes. During the first vaccination campaign 1,736,995 doses of vaccine were administered, and these were followed by a further 1,374,480 doses in the second round, when attention was concentrated on children up to four years, the age group determined most receptive (Koinange, Rogowski and Metselaar, 1973). The authors cited above arrive at the conclusion that under the circumstances at present prevailing in Kenya, an intensive polio vaccination programme of this kind can keep poliomyelitis at an acceptable level in Kenya if carried out at three-year intervals, although rising standards of living and hygiene may well increase the danger of epidemic outbreaks.

2. Viral Hepatitis
(Virus Hepatitis Type A (ICD 070.1) and Virus Hepatitis Type B (ICD 070.2)

Definition: a general illness caused by a virus, the indicative symptom of which is a hepatitis. Hepatitisvirus A (HVA) is regarded as an agent which is transmitted by

faeces (and presumably also by urine and possibly other
secreta, orally as well as parenterally. There is world-wide
endemic occurrence, with frequent lesser or greater epidemic outbreaks. Hepatitisvirus B (HVB) is considered
the agent of viral hepatitis type B, the transmission of
which is as a rule parenteral, although also at times oral.
The evidence of the hepatitis antigen B (HB Ag) makes
it possible to conduct a sero-epidemiological investigation of the distribution of the virus B hepatitis.

No conclusive investigations on the distribution of
hepatitis in Kenya exist (see Fig. 17, back of Map 9). The
distribution of the hepatitis B-antigen among healthy
people in Kenya has been described by Bagshawe and
Cameron (1974). In 3,763 examinations in Mombasa,
Nairobi, Nakuru, Nyeri, Rumuruti, and at Kiteta
(Machakos), 4.3 – 11.9% carriers of hepatitis B-antigen
were found. Whereas the percentage in the lowlands
below 1,500 m did not exceed 5%, it rose together with
increasing altitude. No explanation of this feature was
available, however. The Annual Return of Diseases of
the hospitals and the Weekly Notifications of Infectious
Diseases of the districts (see Fig. 17, back of Map 9), as
far as they cover the central and western areas of Kenya
and the south eastern parts of Uganda were first analyzed
by Diesfeld in 1969 (Diesfeld, 1969 a) and compared
with the figures for poliomyelitis. This demonstrated
that both the infections experienced their widest distribution in the highlands and on the densely settled
slopes of the Aberdares and of Mt. Kenya. There was also
a striking correlation between the figures reported for
poliomyelitis and hepatitis, which may in this respect be
regarded as diseases indicative of inadequate sanitary conditions under circumstances of advanced socio-economic
development and overpopulation.

An analysis of the Weekly Notifications of Infectious Diseases at district level in comparison with the
Hospital Recording Rate of 54 government hospitals reveals that the distribution pattern of clinically-diagnosed
hepatitis is expressed as the average over the last decade.
By far the highest figures occurred in the Coast Province
(40.75/100,000) with particularly high figures appearing
for the hospitals at Msambweni, Garissa, and Kilifi. The
returns for all the other provinces—namely, Central Province, Nairobi City, and North Eastern Province—follow in this rank order at intervals of the power of ten.
The Rift Valley, Eastern, Western, and Nyanza provinces
make up the lower end of the scale.

3. Typhoid Fever and Other Enteritic Infections
(ICD 002 – 003)

In the clinical and epidemiological sense typhoid
fever and other salmonella infections are often grouped
together as enteric fever, although a distinction is possible by means of precise analysis, and—above all—by
bacteriological and serological control. Apart from this,
attention should be given to the numerous shigelloses,

entero-viral infections, amoebic dysentery, cholera, food
intoxication, and intestinal anthrax, all of which can
cause diarrhoeal diseases of varying degrees of seriousness,
accompanied by more or less serious general health conditions.

In Kenya a bacteriological and serological identification can only be undertaken in exceptional cases, at the
Kenyatta National Hospital for example, where cases
were analysed in the 1961 – 1972 period. In this the most
varied serotypes of shigellae, salmonellae, and pathogenic
Escherichia coli were found (Wamola et al., 1974). Definitive statements on this occurrence in Kenya and on
the natural history of the infection cannot be based on
this, however.

Fendall and Grounds (1965 b) have provided a short
review of the state of information concerning the situation in Kenya up to 1963. According to their review,
considerations must start out from the assumption that
typhoid fever in particular is endemic to a high degree,
that the infection of the infant population takes place
relatively early, and that massive epidemic outbreaks are
thus relatively rare. These are in each case prompted by
special local circumstances, resulting in a high number of
cases, which in turn produce particularly high figures for

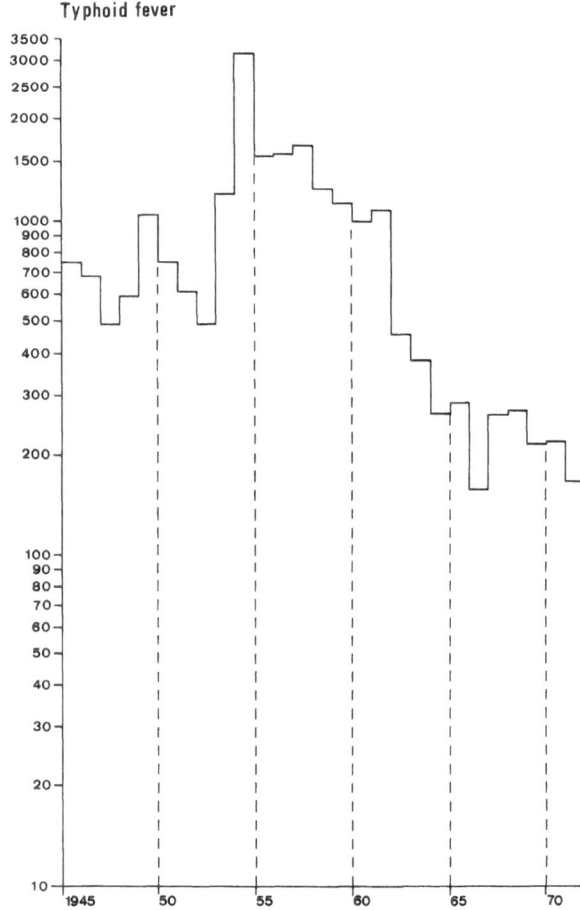

Figure 13. Typhoid fever, number of annual notifications 1945 – 1972
(Source: Fendall and Grounds, 1965 b, Bulletin of Infectious Diseases,
Ministry of Health, Nairobi, 1962 – 1972)

the country as a whole in certain years. However, these do not permit the drawing of any conclusions as to the overall situation in the country.

Until about ten years ago the number of reported cases was especially high in the Central Province. It is not clear whether this is to be attributed to the high number of cases or to an efficient recording service. It appears to be established that an overall declining trend can be observed, which is probably justifiably attributed to the gradual improvement in the general conditions of hygiene and of water supplies (see Fig. 13). The consolidation of the population and the already densely settled urban and peri-urban areas with quite certainly inadequate hygienic conditions do not, on the other hand, exclude the possibility of the general latent infection remaining so high that epidemic outbreaks seldom erupt.

Numerous salmonella and shigella infections arise from unhygienic conditions in the food trades catering for the great majority of the population. Judged by experience so far, the tourist industry, by contrast, seems to maintain a considerably higher standard.

Infantile gastroenteritis, especially linked with malnutrition, still represents 60 – 70% of all deaths in these age groups, as well as more than 80% of all deaths caused by gastroenteritis. Together with respiratory illnesses, this group of diseases occupies first place in the mortality statistics. Every year tens of thousands of out-patients and thousands of hospital patients receive treatment in Kenya. Nothing but a general rise in the standard of living and in the standard of health education, personal hygiene, water supply, and sewage disposal, as well as in food hygiene, will alter this. These measures would do more to improve the nation's health than all medical measures taken together.

4. Cholera
(ICD 001)

In the years 1971 – 1973, cholera occurred in Kenya during the course of the VIIIth pandemic for the first time in this century. Presumably it was imported by nomadic herdsmen from the countries bordering in the north and north east and spread southward. Mngola (1974) has provided a detailed review. The main districts to be affected were Mandera, Wajir, and Garissa in January 1971; these were followed by Galole (Hola), Moyale, and Marsabit in summer of 1971. From Garissa, cholera penetrated as far as the northern part of Kitui District (March, 1971). In March, 1973, cholera once again advanced via Lokitaung up to Samburu in the Turkana District.

Typically the individual cases occurred as acute ones. The overall case fatality rate reached 17%. It has not really been established whether it was the protective measures, among others the aid of the Flying Doctor Service and the "Cordon sanitaire", which stopped the dissemination, or whether the disease collapsed on its own. It was mainly of the Cholera El Tor-Inaba strains (Radojcic, 1973).

5. Amoebic Dysentery
(ICD 006)

Although it is generally accepted that in contrast to the Highlands of Ethiopia (Diesfeld, 1965) amoebiasis does not play a special role in Kenya, stool examinations from Nairobi schoolchildren carried out by Wijers, Kinyanjui, and Rijpstra (1972) produced intestinal protozoa in 75.6% of the cases. Besides *E. coli* and *Endolimax nana,* 15.7% also contained *Entamoeba histolytica.* Indications as to pathogenicity were not detected during the course of these investigations.

Table 32. Hospital Recording Rate (HRR) for "amoebic dysentery" 1962 – 1973 (cases per 100,000 inhabitants in the catchment area as a yearly average)

Kerugoya	519	Embu	259	Thomson's Falls	180
Kiambu	362	Bungoma	231	Muranga	140
Kericho	349	Wesu	209	Meru	131
				Marsabit	129

The analysis of the Annual Return of Diseases for the period 1962 – 1973 showed that the 12 hospitals with the most frequent entries under the diagnosis "amoebic dysentery" had their catchment areas located in the uplands.

By contrast, all the others apart from Galole (127) remained markedly below these levels or reported no amoebic dysentery at all.

6. Tetanus
(ICD 037; 670)

Tetanus, a disease which takes an acute course is caused by the toxin of *Clostridium tetani,* the anaerobic spore-forming soil bacillus. Infection usually occurs by way of open wounds with germ- or spore-bearing material like dirt or filth.

In Kenya tetanus occupies the fourth place in the list of recorded causes of death; this is not so much because of its frequency as because of its high lethality (Ministry of Health, Bulletin of Infectious Diseases 1972). The lethality in Kenya fluctuates between 30 – 50% (Muller, 1974).

In Kenya, the enrichment of soil with tetanus spores is of great significance; they are taken up together with the grass and returned with the cattle dung. The contact with man appears to be particularly close in the case of mixed farming.

According to Rigby (1960), the increasing application of dung in agriculture is largely responsible for the 21-fold increase in the number of hospital admissions of tetanus-infected persons which occurred during the period 1933 – 1957. This increase is far in excess of the population increase or in the popularity of Western medicine and hospitals during the same period. Tetanus is particularly widely found in banana plantations and those of mixed banana and coffee cropping where the technique of mulching favours the clostridae (Diesfeld,

1969 a). Fendall and Grounds (1965 a) had already drawn attention to the correlation between agricultural activity in densely settled areas and the frequency of tetanus occurrence.

Baker (1963) noted the connection between the frequency among female patients in the 20 – 34 years age group with their agricultural activity, in the course of which numerous minor lesions provide entry points for the germs.

Tetanus is well known among the population; in the north of Kenya, where the authorities are probably far less often informed of its presence than in other remaining areas of Kenya, men have their lower incisors broken off as a "preventive measure" to ensure the intake of liquid by means of a straw in the event of lock-jaw (Vanek, 1968, personal comm.).

There are a multitude of cultural and ecological factors which determine the frequency and pattern of tetanus distribution.

A special problem occurs in tetanus in new-born babies (neonatal tetanus), when the end of the navel cord is infected as a result of unhygienic care of the navel at the time of delivery. Above all it is the traditional sealing of the navel with a mixture of earth, cow-dung and ash, widely practiced in the coastal region, which is responsible for countless deaths (Fendall and Grounds, 1965 a).

The number of cases recorded in the "Notifications of Infectious Diseases" 1962 – 1973 in relation to the average district population (Census 1969) (Table 33), as well as the Hospital Recording Rate per 100,000 inhabitants in the catchment area for the period 1958 – 1973 derived from the Annual Returns of Diseases from the 54 government hospitals (Table 34) show the distribution of recorded or hospitalised cases of tetanus. The focus is clearly situated in the Coast, the Central provinces, and in Kericho.

Table 33. Tetanus in Kenya (from: Notification of Infectious Diseases 1962 – 1973) *related to 100,000 inhabitants*

Provinces:			
Coast	23.0	Western	2.6
Central	12.0	Rift Valley	1.8
Nyanza	5.8	Nairobi	1.4
Eastern	3.6	N.E.	0

Table 34. Tetanus in Kenya: Mean annual Hospital Recording Rate (1958 – 1973) (per 100,000 population in catchment area)

upper 40 percentile

Malindi	102	Msambweni	37	Lamu	21
Marsabit	54	Kerugoya	36	Meru	20
Nyeri	53	Kericho	35	Murang'a	20
Kilifi	52	Thika	35	Isiolo	19
Taveta	45	Embu	23	Tambach	17
		Kiambu	21	Galole	16
				Garissa	15
				Makindu	15
				Kisii	14
				Wesu	14

Within the framework of the recently improved ante-natal and child welfare clinics, an attempt is being made to induce a larger number of persons to avail themselves of tetanus prophylaxis by vaccination.

Since only very few women attend ante-natal clinics so far, the problem remains of how to provide pregnant women with an ante-natal tetanus prophylaxis during the third trimenon. The solution to the problem of neonatal tetanus is to be found in combined efforts towards comprehensive ante-natal care, together with health education, tetanus vaccination, and in the training of village midwives. The effectiveness of infant vaccination mostly fails on account of the world-wide and generally familiar high rate of defaulters on the booster immunisation required in tetanus prophylaxis.

IV. Helminthic Infections Associated with Soil, Water, and Food

1. Intestinal Helminths

(With particular regard to Roundworm (*Ascariasis,* ICD 127.0) Hookworm (*Ancylostomiasis,* ICD 126) and Beef Tapeworm (*Taeniasis,* ICD 123.3) infections)

Many intestinal worm infections are distributed throughout the world, some predominantly and others exclusively confined to tropical areas. Furthermore, some of these latter ones occur more or less frequently in all tropical climatic zones, whereas others are known to exist only in certain areas. Although it must be assumed that the majority of the population houses one or other intestinal parasite in the course of his life, its actual significance as regards mortality and morbidity is not quite clear. The intestinal helminths most frequently registered in Kenya are infection with roundworm (ascariasis), hookworm (ancylostomiasis) and beef tapeworm (taeniasis). On account of their differing development cycles, the three species of worms are in some way indicative of different ecological conditions favourable to their distribution (see Fig. 16, back of Map 7). Also fairly distributed are whipworm (*Trichuris trichiura,* ICD 127.3), threadworm (*Enterobius (Oxyuris) vermicularis,* ICD 127.4) and *Strongyloides stercoralis* (ICD 127.2). A not unimportant role is played by *Trichinella spiralis* as a human infection (trichinosis, ICD 124), as well as among carrion-eating carnivores and wild — though not domestic — pigs (Forrester, Nelson and Sander, 1961; Nelson, Guggisberg and Mukundi, 1963; Nelson, Blackie and Mukundi, 1966). With the exception of the analysis of laboratory reports from 43 hospitals compiled over the period 1958 – 1965 (Diesfeld, 1969 b, 1969 c, 1970 b), no thorough epidemiological investigations have taken place.

An analysis of these data in relation to environmental conditions in the hospital catchment areas resulted in increased findings of ascariasis, paralleled by the increase in the population density and the precipitation totals;

population density and precipitation were also closely correlated themselves, a fact that renders interpretation more difficult.

Since in the tropics ascariasis is essentially a dirt and household infection, spreading in the immediate surroundings of dwellings, increasing settlement density and subsequent decline in standards of hygiene are believed to promote the possibilities of distribution of the very resistant ova of *Ascaris lumbricoides*. Hookworm diagnoses were found to decrease with increasing altitude of the hospital catchment areas. A more detailed analysis including the annual climate, defined as "temperature-humidity-milieu" (Diesfeld, 1970 b), in the hospital catchment area demonstrated, by contrast with ascariasis, a regression in the infection frequency at both cool and dry (i.e., higher) and at warm and dry altitudes, whereas warm and humid altitudes did not produce any significant differences in the diagnostic frequency of ancylostomiasis and ascariasis. The parasitological identification shows *Necator americanus* as a rule, whereas *Ancylostoma duodenale* is only found in the coastal region (Foy and Kondi, 1960). Hookworm infection is highest in the coastal region and in the vicinity of Lake Victoria; in these zones hookworm plays a special role as the cause of serious anaemias.

The frequency of taeniasis diagnosis at the hospital laboratories confirmed the expected connection with the cattle population in the hospital catchment areas, together with the simultaneous decrease of the remaining helminths, especially of ancylostomiasis. *Cysticercus bovis,* the larval stage of the beef tapeworm, is found in about one-fifth of all lifestock. *Taenia solium,* the pork tapeworm, causes cysticercosis in pigs, but has so far not been found in man. *Hymenolepis nana,* the dwarf tapeworm, is often detected in children, but seems to be of little significance in Kenya from a clinical point of view (Rees, Mngola, O'Gary and Pamba, 1974). Cestodes, in which man acts as an intermediate host, and in particular the hydatid disease, will be discussed in another chapter.

The regional distribution of the diagnosed frequency of ascaris, ancylostoma, taenia, and schistosoma can be presented in a geomedical analysis based on the cumulative data of the "Annual Laboratory Returns" from the period 1958 – 1973; these refer to 54 hospitals in Kenya with a total of 321 laboratory/years, an average per hospital of 6.4 years, and over 1.1 million stool examinations and half a million urine tests (see Map 8). As already demonstrated in an earlier analysis of this kind (Diesfeld, 1969 a), this extended data "collective" permits collation of the individual places of investigation into groups having similar distribution patterns of helminths and similar geographical, ecological, and epidemiological conditions (Table 35).

Typical patterns of helminth distribution can be observed as indicative of these conditions, if due care is taken in the interpretation (Table XI, Annex).

The following 9 types can be distinguished:

1. *East Coast Type:* characterised by the dominance of hookworm and *Schistosoma haematobium* frequencies in diagnosis and the virtual absence of *S. mansoni* and taeniasis. Although it is located on the coast, the urban Lamu does not fit into the picture, partly because of the high proportion of *ascaris,* the almost complete lack of ancylostomiasis, and partly because the helminth pattern classes it together with Nairobi in an area of higher altitude and denser settlement. Galole, situated on the middle course of the Tana River and characterised by intensive irrigation schemes, must, on the other hand, be included in the coastal region.

2. *Kamba Type:* The Kamba type in Machakos and Kitui is characterised by the occurrence of both the agents of schistosomiasis: *Schistosoma mansoni* and *S. haematobium,* as well as a large proportion of taenia and a small proportion of *ancylostoma* by comparison with the frequency of *ascaris* diagnosis.

3. *Aberdare Type:* The Aberdare type represented by hospitals from the south eastern and eastern outliers of the Aberdares to the irrigation areas of the Mwea Plains, as well as the ecologically and agro-geographically similar areas around Voi, Taveta, and Makindu, is characterised by a roughly equally high proportion of *ascaris* and *ancylostoma* findings, together with a not unimportant proportion of *S. mansoni. S. haematobium,* however, occurs only on a small scale in Makindu and Voi, but to a considerable degree in Taveta.

Table 35. Relative frequency of Helminth-Laboratory findings from 54 "Annual Laboratory Returns" of Kenya Hospital Laboratories (1958 – 1973), grouped in 9 geo-ecological types

Geoecological Type	Ascaris	Ancylost.	Taenia	S. mansoni	S. haematob. *
East Coast Type	31.2	65.8	2.2	0.5	28.10
Kamba Type	35.5	11.3	10.5	42.5	10.83
Aberdare Type	44.1	32.6	14.5	7.9	5.97
Nyeri-Mt. Kenya Type	72.4	14.3	11.4	1.7	3.38
Narok-N.E. Type	37.9	12.7	42.5	6.9	0.51
Rift-Valley Type	37.7	14.2	47.6	0.5	0.36
Kisii Type	51.5	12.2	27.2	2.5	1.67
Nyanza Type	47.9	26.9	20.3	14.6	7.75
W. Province Type	27.1	51.1	18.4	0.3	0.32

* S. haematobium: = % pos of urine examination

4. *Nyeri-Mt. Kenya Type:* This type is characterised by a particularly large proportion of ascaris diagnosis and a very small proportion of *ancylostoma* diagnosis. Those hospitals situated at the middle and higher settlement altitudes around Mt. Kenya and those in the northern Aberdares present a striking homogeneity in their helminth pattern as they do in the ecological conditions of their catchment areas. The coastal town of Lamu and the hospital at Wesu, which is situated on a mountain-massif, can also be classified with this type. Both these hospitals, together with those of Embu and Naivasha, still manifest a high proportion of *S. haematobium* diagnoses; Naivasha's high proportion is striking, but it is not substantiated in any other way. The high number of findings in Wesu, Lamu, and Embu are the outcome of the adjacent endemic areas of *S. haematobium.*

5. *Narok-North-eastern Type* and

6. *Rift Valley Type:* As with those of northern Kenya, the hospitals of the northern, central, and southern Rift Valley show an absolute preponderance of findings of *taenia* and, by comparison with the much reduced findings of *ancylostoma,* a large proportion of *ascaris* diagnoses. The difference between both types consists of a small proportion of *S. mansoni* still to be observed in the southern Rift Valley and in north eastern Kenya, which occur only very sporadically in the remainder of the Rift Valley in northern Kenya, and never autochthonous. They result from importations by patients from outside. Both of these areas are characterised by extensive semi-nomadic and nomadic animal husbandry and large-scale farming, and by no more than marginal subsistence agriculture in favourable localities.

7. *Kisii Type:* Helminthologically this type is characterised by a 3 : 2 : 1 relationship of *ascaris : taenia : hook-worm,* with findings of *S. mansoni* and *S. haematobium* occurring only sporadically. This rather homogeneous pattern of helminth distribution applies to the hospitals at the higher altitudes of Kenya's western highlands, where agriculture tends to be mixed. A similar pattern can be observed at Nakuru, which has a very large catchment area, and, interestingly, at Garissa.

8. *Nyanza Type:* Kisumu and Homa Bay, situated on the shores of Lake Victoria, are characterised by a proportion of *ascaris* findings among the intestinal helminths on the whole twice as large as those of hookworm and with only a small proportion of *taenia.* The relatively significant proportion of *S. mansoni* and *S. haematobium,* both of which are characteristic of the coast and the coastal hinterland of Lake Victoria, is conspicuous.

9. *Western Province Type:* In a manner similar to that of the East Coast, hookworm predominates absolutely over *ascaris,* with only relatively small numbers of *taenia* and an almost total absence of *schistosoma* findings.

2. Schistosomiasis
(ICD 120)

Schistosomiasis is a chronic endemic disease caused by a trematode of the genus *Schistosoma.* The intermediate host is a snail and the definitive host is men or animals. The free-swimming cercariae are discharged by the snail, pierce the skin of men, and reach the portal venous system where they develop into adult worms which migrate to the smaller veins of the bladder in the case of a *Schistosoma haematobium* infection or to the mesenteric veins in the cases of a *Schistosoma mansoni* or *Schistosoma japonicum* infection. In these sites the female deposits eggs that are excreted in the urine or stools. It is mainly the eggs which are responsible for the clinical features of the disease.

a) *The distribution of schistosomiasis in Kenya:* As everywhere in Africa, so in Kenya schistosomiasis occurs in the form of foci (see Map 9). Areas highly infested with one or both species of *Schistosoma mansoni* or *Schistosoma haematobium* alternate with others free of schistosomiasis. Besides the ecology of the transmitting snails, which essentially determines the distribution, it is the increase in irrigated agriculture of different kinds and the close contact between man and water which determines the pattern of distribution.

Apart from increased information on the distribution, which seems to indicate an increase in the problem, the increase in the number of small- and large-scale irrigation schemes does result in an authentic increase in the number of infected persons in these areas. At the present time control measures can be carried out only in large-scale irrigation schemes which are centrally managed and centrally supplied with water. The great majority of individual contact possibilities during the parasite's final development cycle remain largely unavoidable at the level of the smallholding, as long as the appropriate health education measures are not carried out and made effective. Schistosomiasis is one of those parasitic mass infections in the tropical developing countries the long-term control of which depends more on the improvement of the general conditions of living, hygiene, water, and sewage disposal and an appropriate health education and irrigation planning than on the application of available chemo-therapeutical means for the elimination of agents in man and the extermination of the molluscs acting as intermediate hosts.

Scarcely any agricultural irrigation scheme in an area of potential transmission has succeeded in avoiding the infection of the population by schistosomiasis through appropriate planning or by interruption of transmission in an already-infected population. However, one such model is to be found in Kenya in the Mwea Tebere Irrigation Scheme, when after 15 years' duration and progressive infection schistosomiasis was finally brought under control. Not only knowledge, but also the implementation of adequate measures resulting from the knowledge of the ecology of the final and intermediate hosts are required if this chronic disease is to be exterminated.

b) *The ecology of the intermediate hosts in Kenya*

aa) *Intermediate hosts of S. haematobium* (after Highton, 1974 b):

The principal intermediate hosts of *S. haematobium* in Kenya are snails of the *Bulinus* genus and the subgenus *Physopsis,* which belong to the *Bulinus africanus* group. They occur extensively at altitudes below 1,800 m above sea-level. Above this height the water temperatures are too low for the optimal breeding conditions of *Bulinus.* Its habitat is generally to be found in perennial as well as seasonal streams, especially in dammed waters, and also in canals, furrows, and depressions containing water, reservoirs, drainage and irrigation ditches etc. The snails can survive 5 to 8 months of dehydration during the summer months ("Estivation" according to Webbe and Msangi, 1958).

B. (P.) africanus occurs from the coastal area up to altitudes of 1,800 m; at the higher altitudes it is chiefly found in Kitui, Machakos, Kiambu, and Murang'a districts. *B. (P.) globosus* is the chief vector in the coastal plains and in the valley of the Tana River. It rarely occurs above 1,250 m. *B. (P.) nasutus* occurs sporadically in isolated foci. *B. (P.) nasutus productus* is mainly responsible for the transmission of *Schistosoma haematobium* in the Nyanza and Western Provinces.

bb) *Intermediate hosts of S. mansoni* (after Highton, 1974 b): All species of the genus *Biomphalaria* are receptive to infections with miracidia of *S. mansoni,* and are thus also potential vectors. *Biomphalaria pfeifferi* and *B. sudanica* occur in a wide range of water accumulations —puddles, reservoirs, dams, ditches, canals, drainage ditches and even concrete-lined reservoirs—but seldom in swamps. *B. sudanica* occurs chiefly on the shores of lakes and swamps. It is of interest that species of *Biomphalaria* have never been found in the coastal region but only over 300 m above sea-level (Teesdale, 1954, 1962; Teesdale and Nelson, 1958). As a result of similar observations in other regions of East and South Africa, Sturrock (1965) tends to believe that water temperatures in the coastal hinterland of the Indian Ocean are too high for optimal living conditions of *Biomphalaria.* The absence of *Biomphalaria* is confirmed by the absence of autochthonous *S. mansoni* infections in these zones.

c) *Regional distribution of the frequency of schistosomiasis in Kenya:* Apart from investigations in the organised irrigation schemes, little information was available on the real distribution of schistosomiasis among the population until a few years ago. In the meantime it has been estimated that about 1 million inhabitants of Kenya are infected by these parasites. The hospital laboratory reports were consulted in order to gain a general view of the whole country. However, it proved impossible to draw any far-reaching conclusions from this concerning

the actual state of infection. A summary and analysis of Annual Laboratory Returns from 54 hospital laboratories over the period 1958 – 1973 showed 1 million stool examinations and almost half a million urine tests for 321 laboratory/years (see Table XII, Annex). A regional distribution of the frequency of *S. mansoni*—positive stool tests—and *S. haematobium*—positive urine tests—clearly shows the previously known distribution pattern of the transmitting snails as well as the probability of autochthonous transmission and transmission-free areas (see Map 8). Together with the investigations recently carried out by the Parasitological Department of the Division of Vector-borne Diseases of the Ministry of Health and references in the literature it results in the following picture:

1. *East Coast and Tana River Valley:* On the coastal plain and in the lower and middle Tana River valley *S. haematobium* is the only form occurring. Stool examinations at Shimba Hill (Kwale District) produced egg excretion in 22.4% of all urine tests; in Kilifi this rose to 67.5% of all cases (DVBD Coast Province, Annual Report, 1972). In the Hola irrigation schemes (Galole on the Tana River), which comprise some 800 ha. of irrigated land, at present mechanical snail control is carried out, albeit so far without any recognisable reduction in the infection rate among the population. Even prior to its implementation it was possible to note a decrease in infection of the population proportional to the distance from the river (Teesdale, personal communication cited by Highton, 1974 b, op. cit.).

45 km north of Hola in Bura a further irrigation scheme is planned which will assuredly present an important agricultural reserve but will also embrace the problem of bilharzia from the very outset.

S. mansoni is practically non-existent in the coastal region. The same applies to those species of *Biomphalaria* acting as intermediate hosts.

This picture is reflected in the Annual Laboratory Returns of 1958 – 1973 (see Table 36 and 37). Between 14 – 45% of all urine tests are positive, whereas the stool tests are practically free from *S. mansoni*.

2. *Taveta:* There is a further important focus of *S. haematobium,* mentioned by Fendall and Grounds (1965 b), in the sisal plantations of Taveta on the border

Table 36. Stanine transformation of stool tests Annual Laboratory Returns 1958 – 1972 for S. mansoni

<1.0%	1 – 2%	2 – 3%	3 – 9%	>9%
	Mandera	Gatundo	Kimbimbi	Kangundo
	Kiambu	Thika	Kitui	
	Kisii Nandi	Taveta	Machakos	
	Kajiado	Makindu		
	Voi, Wajir	Londiani		
	Homabay			
	Kisumu	12%	7%	4%
60%	17%		23%	=100%

Table 37. Stanine transformation of urine tests Annual Laboratory Returns 1958 – 1972 for S. haematobium *

<0.5%	0.5 – 3%	3 – 10%	10 – 30%	30 – 45%
Maralal, Kapsabet	Busia, Nakuru, Kapsabet	Voi	Kilifi	Kinango
Moyale, Tambach	Th. Falls, Machakos	Kisumu	Mombasa	Kipini
Murang'a Kimbimbi	Kericho, Kisii, Thika	Mandera	Malindi	Galole
Kianyaga, Kakamega	Wajir, Embu, Makindu	Nandi	Msambweni	Taveta
Bungoma, Nanyuki	Londiani	Wesu	Kitui	
Kitale, Narok, Eldoret		Kangundo	Garissa	
Nyeri, Alupe, Kabarnet		Lamu	Homa Bay	
Kiambu, Kajiado			12%	7%
Kerugoya		17%		19
36%	20%		36%	

* The presence of *S. haematobium* in the urine sediment is determined by a concentration procedure. Moreover, *S. haematobium* has a significantly higher egg production, which results in the proof of *S. haematobium* being much more sensitive, comparatively speaking, than that of *S. mansoni* in a simple stool smear. In both cases, however, analysing the laboratory data the stanine transformation of frequencies emphasises the most important endemic areas.

with Tanzania to the east of Mt. Kilimanjaro. The Annual Laboratory Returns show a positive diagnosis in 45% of all urine tests. Although the proportion of *S. mansoni* amounts to only 8% of the helminth-positive stool tests, a much greater infection rate must be assumed, since the appropriate intermediate host occurs prolifically in the 4,000 ha. irrigation area (Chowdhury, 1975). The contiguous districts of Wesu and Voi show only a relatively small proportion of schistosome findings in the case of both *S. mansoni* and *S. haematobium.*

3. *Kamba-Kitui:* A large endemic area predominantly of *S. mansoni* is to be found in the hill country of Machakos District (Machakos, Kangundo, Makindu). By comparison with neighbouring Kitui, *S. haematobium* is relatively sparse. An examination of 7,000 schoolchildren in Machakos township over the period 1961 – 64 established an infection rate of 17% (Teesdale, 1966). Examinations of 2,900 schoolchildren undertaken in six divisions of the district by Mutinga and Ngoka (1971) demonstrated an average infection rate of 29.5% (see Table XIII annex) (Chowdhury, 1975, op. cit.).

3 a) *The adjacent highland around Kitui* constitutes an isolated climato-ecological focus in which examinations of more than 5,000 schoolchildren at 27 schools established the existence of a 17.5% infection rate. The frequency fluctuated between 2.5 – 26.5% (Annual Report DVBD 1972 – 1974). The Annual Laboratory Returns from the Kitui District Hospital reveal a mean frequency of 4.6% of *S. mansoni* diagnosis in stool tests, and a mean frequency around 32.5% of *S. haematobium* in urine tests.

4. *Aberdare and Mwea Tebere Irrigation Scheme:* The oldest and so far the largest irrigation scheme is located in the Mwea Plains in the upper Tana river basin at the foot of the Aberdares and Mt. Kenya. Since 1956 rice has been grown on irrigated land, which now covers an area of 5,500 ha. At present there are more than 3,000 formerly landless settlers living here, a total of about

30,000 persons spread over 34 settlements. Teesdale (1961) noted as recently as 1961 that schistosomiasis did not occur in this region. Nevertheless, as a result of increasing irrigation agriculture, schistosomiasis spread rapidly during the ensuing fifteen years. The settlers from the surrounding areas of Kiambu, Murang'a, Nyeri, Embu and Kirinyaga districts, where the problem of schistosomiasis was not known at this time, evidently brought in *S. mansoni,* which spread rapidly in the irrigation scheme.

The waters joining the irrigation network from the Nyamindi and Thiba rivers brought the snails. While the massive use of molluscicides since 1967 has succeeded in reducing the infestation by *S. mansoni* within the area of the Mwea Tebere Irrigation Scheme—at the Kimbimbi and Kombueni schools to 8.2 and 5% respectively—infection among children in the surrounding divisions remains between 30 – 60%. In Murang'a and Kiambu districts, too, and particularly at lower altitudes the infection rate in schools fluctuates between 6.4 – 38.2%; in this the Gatundu, Kiambaa, and Kijabe divisions stood out especially as foci. The Annual Laboratory Returns from the area's hospitals also reflect this situation.

5. Another endemic schistosomiasis area is the *Nyanza Region* on Lake Victoria. While the population living in Tanzania along the shores of Lake Victoria suffers considerably from schistosomiasis, especially in the area around Mwanza, infection by *S. haematobium,* as well as by *S. mansoni,* is much reduced in spite of the occurrence of transmitting snails. Only on the islands of Mfangano (Wijers and Munanga, 1971) and Rusinga (Pamba, 1974) was *S. mansoni* found—in 46% and 30 – 60%, respectively, of the schoolchildren examined. On Mfangano *Biomphalaria* occurs more in the interior, on Rusinga, more on the coast of the island. *S. haematobium* does not appear to be much in evidence. On the mainland *S. mansoni* prevails among the fishermen along the coast, whereas *S. haematobium* predominates among the farmers in the coastal hinterland. The Annual Labo-

ratory Returns for Kisumu and Homa Bay report *S. haematobium* in 3.85% and 1.7% resp. of all urine tests. In both laboratories *S. mansoni* is reported in 14.6% of stool samples.

The investigations of Kinoti (1971 a, b) conducted among 1,700 schoolchildren on the Kano Plains produced evidence of only 4% *S. haematobium:* this low grade of infection is correlated with the absence of *Bulinus africanus* and *B. nasutus,* the main intermediate hosts on the "black cotton soil" of this area.

The occurrence of *B. globosus* as isolated foci only in the Kano Plains is evidently responsible for the small degree of infection. The importation and establishment of transmitting snails in the train of large-scale, planned irrigation measures, and with it the massive occurrence of *S. haematobium,* can only be avoided by appropriate planning. Outside the Kano Plains *S. haematobium* is transmitted by *B. globosus, B. P. forskali,* and *B. P. africanus* (Brown, 1975).

An extension of the irrigation installations into the northern and eastern fringe areas of the Kano Plains, offering a potential of 14,000 ha., would open up contact to *Biomphalaria pfeifferi*—the highly potent intermediate hosts in that area, which multiply particularly successfully in irrigation installations (Brown, 1975). Periodic flooding and waterfowl present further possibilities for the spread of the snails.

Schistosomiasis is no problem in the *Ahero Pilot Scheme* which comprises an area of 880 ha. lying 22 km east of Kisumu and under control since 1968. It remains to be seen how far this will apply to more extensive areas.

The Yala Swamps, in the Busia and Siaya Districts of western Kenya, of which 6,000 ha. have been reclaimed with the assistance of the United Nations Development Programme (UNDP), is considered suitable for irrigation in the distant future. A 200 ha. pilot project was established at Bunyala. Schistosomiasis, both *S. mansoni* and *S. haematobium,* is already endemic in this area, where people have continuous contact with water. *B. sudanica* is confined to the lakeshore and edges of the swamps, whereas *B. pfeifferi* is common in the irrigation scheme (Chowdhury, 1975). Control measures since 1970/71 have reduced snail breeding.

The only irrigation scheme in which schistosomiasis has so far not been reported is at Perkerra, although the adjacent Lake Baringo harbours *B. sudanica.*

6. In the *Western Province* (Kimilili near Kakamega) the DVBD recently discovered another focus of *S. mansoni* and *S. haematobium,* in contradiction to earlier assumptions, a higher rate of infection is also to be expected here. At altitudes above 1,800 m schistosomiasis does not occur as an autochthonous infection. Autochthonous occurrences of schistosomiasis need not be anticipated either in the arid and semi-arid areas because of the absence of perennial streams, as long as no artificial irrigation is introduced which brings in more stagnant or running water than is the case in irrigation at intervals, for example, at Lake Baringo or in Galole.

A recent investigation by Brown (1975) was the first in Kenya to draw attention to the well-known association between haematobium infection and cancer of the bladder. In contrast to the white population in areas free from bilharziasis, the biopsy material from 76 cases presented peaks in two age groups: namely in the 4th and the 7th decades. Carcinomas of the bladder unambiguously, histologically associated with bilharziasis of the bladder were found in 39% of the cases. The mean age of these patients was about 45 years, that of the remaining patients 60.7 years. The regional distribution of the origin of these patients coincided with the areas in which *S. haematobium* occurs, whereas the occurrence of the remaining carcinomas of the bladder does not show such an indicative frequency.

This geomedical presentation of the problem of schistosomiasis clearly demonstrates the extraordinary problem in Kenya. On the one hand, marked scarcity of land inevitably demands large irrigation schemes, whereas on the other, the indisputably joint problem of the spread of schistosomiasis—that is, its control simultaneously with the development of the scheme—constitutes a considerable financial burden. Even if Highton and Chowdhury (1974) were able to show that snail control programmes so far conducted in the five existing irrigation schemes were relatively reasonable in expenditure in the context of overall economic development, they constitute a considerable, if not an intolerable, financial burden.

V. Anthropozoonoses

The term anthropozoonoses is applied to infectious diseases the agents of which are pathogenic to man and animal alike. Applying the term in a somewhat narrower sense, it refers particularly to all those infections of domestic animals which are of special importance to man as diseases—like anthrax, brucellosis, and echinococcosis (hydatid disease). In a broader sense, even leptospirosis and plague ought to be counted among the anthropozoonoses, but the latter will be discussed in the section of vector-borne diseases.

In Kenya relatively little information is available on the four most important anthropozoonoses mentioned above. It must be assumed that they are of greater significance than is presently accepted.

1. Anthrax
(ICD 022)

Anthrax is an infection of cattle and game by the spore-forming *Bacillus anthracis,* which may under certain circumstances also occur in man as a highly acute, febrile, septicaemic, and often lethal disease. A cutaneous form is transmitted by skin contact with infected animal material, chiefly hair, wool, skins, excreta, as well as with soil and dirt contaminated by them. An intestinal form results from the consumption of infected meat. A pulmonary form in man is transmitted by inhalation of contaminated dust chiefly in those branches of industry concerned with the processing of animal products.

Anthrax seems to occur sporadically among Kenyan livestock, according to observations at abattoirs under the supervision of the Veterinary Department. Two-thirds of the cases originate from the Rift Valley Province, chiefly from the Kipsigis, Narok, Kajiado, and Sirikwa districts; most of the remaining cases come from the Central Province (Annual Report Veterinary Department, 1962 and 1963).

The forms observed among men are predominantly intestinal anthrax as acute diarrhoea. Mass infections following consumption of meat from collapsed or moribund cattle occur frequently and are particularly striking features in connection with local traditional celebrations (Fendall and Ground, 1965 a). As in most cases the diagnosis is only a clinical one and lacks bacteriological confirmation, it is not always possible to distinguish, whether as such it was an anthrax infection or not. Figures taken from the hospital reports agree with the Notifications of Infectious Diseases (see Fig. 14 and Fig. 17, back of Map 9) on the regional distribution of anthrax among domestic animals as indicated by the Veterinary Department. The long-term averages of the "Weekly Notifications of Diseases" present the highest figures reported per 100,000 inhabitant for the Rift Valley Province, which is followed by the Central, Eastern,

Coast, and Nyanza provinces. The remaining provinces either report no cases or only isolated anthrax occurrences in their long-term averages. The same applies to 77% of the 54 government hospitals: only 14 of them register more than 10 cases per 100,000 inhabitants from within their catchment areas as a long-term average and these are almost exclusively located in the provinces mentioned above.

2. Brucellosis
(Undulant fever, ICD 023)

In East Africa brucellosis is an infection of cattle involving *Brucella abortus* and of goats, and sheep involving *Brucella melitensis*. The germs are transmitted through milk and inadequately sterilised milk products, as well as meat, blood, placenta, still-born foetuses and the excreta of infected animals. Its distribution among livestock is not even approximately known. As an infection it constitutes a considerable economic loss, since it leads to miscarriage on an epidemic-like scale (see Fig. 17, back of Map 9). The infection rate of man was found to be particularly high where *B. melitensis* infection of goats and sheep was predominant. The hygienic standard of cattle, goats, and sheep under semi-nomadic or smallholder conditions seems to enhance the spread of brucellosis among animals and between animals and man more than under nomadic conditions (Vanek, 1976). Serological examinations of cattle, goats, and sheep over the period 1969 – 1972 showed a positive agglutination test among 1.5 to 7.5% of the animals tested, with no discernible connection between the type of animal husbandry and the degree of infection.

A remarkably high percentage of 10.6 – 11.1% was found among cattle of the Coastal Province and in Narok. No positive titer was found in the Nyanza Province (Oomen and Wegener, Oomen, 1976 and Table 38).

Although the diagnostic possibilities are limited, the Weekly Notifications of Infectious Diseases as well as the Hospital Reporting Rate of the Annual Return of

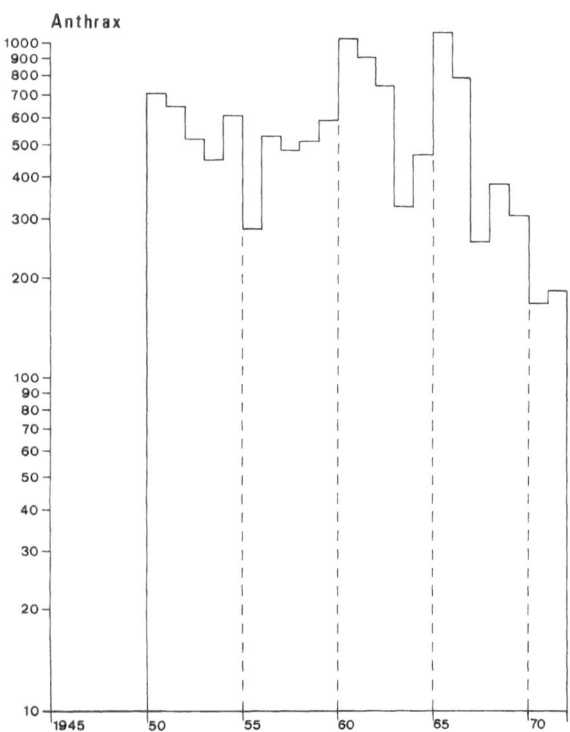

Figure 14. Anthrax, number of annual notifications 1950 – 1972 (Source: Fendall and Grounds, 1965 a, Bulletin of Infectious Diseases, Ministry of Health, Nairobi, 1962 – 1972)

Table 38. *Serological Survey among cattle, goats and sheep in Brucellosis (1969 – 1972)*

Area	Cattle		Goats and Sheep	
	No.	% positive	No.	% positive
Turkana	201	1.5		
Marsabit			187	1.6
Samburu, Isiolo	557	5.2		
North Eastern Province	1,472	7.5		
Baringo			326	3.4
Laikipia	487	2.1		
Narok	667	11.1	807	1.4
Kajiado	557	3.9		
Machakos, Kitui	46	4.3		
Nyanza	185	0		
Kiambu	208	2.4		
Coast Province	724	10.6		

Source: Oomen and Wegener, 1974

Diseases allow a largely coincidental tendency to be observed in an average of ten years (see Fig. 17, back of Map 9). The highest figure is reported from the North Eastern Province with 3.66 per 100,000 inhabitants, followed by the Coast Province with 1.8 per 100,000 inhabitants, the Rift Valley Province with 1.49, the Eastern Province with 0.63, and the Central Province with 0.36. No cases of brucellosis appear in the long-term averages of the remaining provinces. The same thing applies to over 60% of the 54 government hospitals. Of the 21 hospitals which report brucellosis, only six record more than 10 per 100,000 inhabitants in their catchment area. These are the hospitals of Marsabit, Taveta, Voi, Wesu, Narok, and Kajiado (see Fig. 17, back of Map 9), this matches with the findings of Vanek (1976).

3. Echinococcosis
(Hydatid disease, ICD 122)

Echinococcosis is the term applied to the infection of man by the larvae of the canine tapeworm *Echinococcus alveolaris* and *Echinococcus cysticus,* the adult form of which lives as a parasite in the intestines of canines such as dogs and foxes. Generally cattle, sheep, and other mammals act as intermediate hosts, although man does so accidentally. The development cycle is closed when canines eat material containing cysts; adult worms develop in the intestines of the canines, and eggs are passed with the faeces, which in turn are taken up orally by the intermediate host on the pasture. Man is infected directly or indirectly by close and unhygienic contact with infected dogs and with dog faeces. Cystic echinococcosis form inside man when he is infected by *Echinococcus granulosus* and the alveolary echinococcosis when infected by *Echinococcus multilocularis.* The analogous infection of cattle, sheep, and goats occurs—so far as this is known through veterinary investigations—in up to 30% of all livestock in East Africa (Fendall and Grounds, 1965 b). Echinococcosis often occurs in association with brucellosis and anthrax in nomadic cattle economies where many dogs are kept which live in close contact with the people (Bell, 1956; Nelson and Rausch, 1963). The highest level of distribution of echinococcosis is found among the Turkana of northern Kenya, a fact which is also reflected in the hospital recording rates. The only hospitals recording a long-term average of echinococcosis are: Lodwar (116 cases per 10,000 inhabitants in the catchment area), Lokitauung (48), Marsabit (10), Narok (10), and Nakuru (2). In the remote hospitals of the north it is especially the surgeons of the Flying Doctor Service, who have specialized in operations on what are often gigantic hydatid cysts.

Recently O'Leary (1976) pointed again to the fact that hydatid disease is mainly concentrated in the northern parts of Turkana district. The mode of transmission is through close contact between infected dogs and infants within the homesteads. This is also reflected in the age and sex distribution: 42% of all hydatid cysts seen

were in children under the age of 15, 55% of the children were females, while in adults 64% were females. Since the dead bodies, except those of the grandparents, are not buried and are therefore consumed by wild animals, man is not a dead-end host.

4. Leptospirosis
(ICD 100)

Another anthropozoonosis, leptospirosis, affects mainly rodents and domestic animals. Human infection occurs accidentally by direct contact with infected animals or through water which has been contaminated by sick animals. The causative agents are different, serologically discernible types of *Leptospira interrogans.*

In Kenya leptospirosis was first discovered among dogs by Piercy (1951). It was first described by Burdin et al. in 1958 in the course of observing an outbreak among cattle, sheep, and goats in Nanyuki District that had been caused by *Leptospira grippotyphosa.* As far as is known to date, leptospirosis in man in Kenya occurs in foci; labourers on the sugar plantations appear to be particularly prone to infection, to such a degree that it might be called an occupational disease. In the vicinity of Kisumu (Nyanza Province) on the sugar plantations of the Kano Plains 24% of the healthy persons medically examined produced positive serum titer; in the vicinity of the Shimba Hills and in the sugar cane plantations at Ramisi, south of Mombasa, the level rose to 26%. In accordance with the degree of exposure at work on the sugar plantations, it is young men who are particularly prone to infection. The causative agents belong mainly to the serum groups *L. australis, L. ictero haemorrhagiae,* and *L. hebdomadis.*

In the sugar plantations it is the bush rat *(Aethomys kaiseri)* which may be regarded as the main reservoir of *L. hebdomadis,* since it displays a 5.7% infection rate. Cattle seem to be by far the most important reservoir (Ball, 1966). The desert gerbils *(Tatera robusta* and *Acomys wilsoni)* also manifest high infection rates, fluctuating between 9.7 – 6.2%; they are the most important reservoir of leptospiroses of the *Icterohaemorrhagiae* serogroup. No reservoir of *L. australis* has so far been found (Forester et al., 1969; Kranendonk et al., 1967, 1968; De Geus et al., 1969; De Geus, 1974).

VI. Nutritional Problems and Anaemias

1. Nutritional Problems

In Kenya, as in many other African countries, the nutrition of the population, especially that of children, and all the problems connected with it must be considered under two aspects: the ecological "productive" and the socio-economic "consumptive". Although these two aspects are frequently inter-related, they are in themselves affected by other factors. The problems relating to ecology and production have been described in Chapter A. IV, while those concerning decisions on production and distribution have been referred to on more than one occasion.

To sum up, it may be stated that ever since the colonial period it has been the declared aim of agricultural policy to give priority to food supplies for the population, particularly that affecting the self-sufficiency in home-grown products of the farming community. It was not until the basis of guaranteed supplies for home consumption among the farming population was secured that the commercialisation of small-scale African agriculture, i.e., cash cropping, was propagated. During the colonial period the development of the large farms remained, almost to the end, in the foreground of agricultural policy; small-scale agriculture began to secure increasing support and encouragement in the move from a purely subsistence economy to one partially based on market orientation after the implementation of the Swynnerton Plan.

In the pre-colonial and early colonial times, crop failures resulting from droughts and vermin attack, possibly combined with epidemics, led to famine and population losses. Events of this sort are to be found in reports for 1890, 1907, 1919, 1943, 1961, 1963, 1966, and 1971/72. Apart from these, the unpublished agricultural gazetteers compiled at district level by administrative officers in the late fifties tell of numerous regional crop failures and crises in food supply. As a rule, the stocks of foodstuffs produced on the farm only last from one harvest to the next. Thanks to the construction of facilities for transport, marketing, and storage, Kenya is now able by its own resources to meet bottlenecks in supply occasioned by regional crop failures and to even-out the considerable differences in yields that occur from one year to the next.

However, in the densely settled highlands the nutritional status of farmers with little or no land is deteriorating as the increase in population reduces the proportional availability of land. The reduction in soil fertility due to over-cropping, which has been described above, is further aggravating the position of this group of the population. If a crisis in the food supply for the rapidly increasing population is to be avoided, the stepping up of food production within the country will continue to be the most urgent task in the formulation of Kenya's agricultural policy. Improvement of cultivation methods, selection of seeds, use of fertilizers and the extension of irrigated cultivation can be employed in further increasing the productivity of large and small-scale agriculture. But the actual problem of nutrition is not only a question of improvements in agricultural techniques. It arises in the socio-economic and socio-cultural sectors as well—as has been pointed out time and time again in the annual reports of the provincial medical officers and of the Ministry of Health. The Annual Report of the Ministry of Health for 1962

(Annual Report, Min. of Health 1962) refers to the increasing problem of protein and calorie malnutrition (PCM). Besides drawing attention to the tradition-conditioned deficiencies and nutritional insufficiencies arising from too short births-intervals and ignorance of a more suited weanling diet, there is also special emphasis on nutritional disorders resulting from intensified food production for the market at the expense of that for domestic requirements.

A first Nutrition Survey, covering the period 1964 – 68, was carried out by the Ministry of Health with the support of the WHO, FAO, and UNICEF. On a sample basis this provided important information on the nutritional condition of the population (M. Bohdal, N. E. Gibbs, W. K. Simmons, Nairobi 1969: mimeographed report to the Government of Kenya). Three sub-locations of the Kikuyu tribe, two of the Luo, and two of the Kamba were selected for the investigation of the nutritional conditions as well as the dietary habits of the population. Though it had become more secure with regard to the regularity and quantity of yields since the introduction of maize as the staple crop, the traditional and formerly well-balanced diet has become unbalanced in respect of both amino acids and mineral content.

The Nutrition Survey showed that in the three areas under investigation, 75.8% of the calorie intake was contributed by carbohydrates, 14% by protein, and 10% by fats. Samples taken in the Central, Nyanza, and Kamba regions during the 1965 period of the investigation revealed that towards the end of the agricultural year more than 50% of the families consumed less than 80% of the appropriate quantity of foodstuffs. The individual supply situation varies according to the season; it is precarious before the harvest, and on the whole is always threatened by recurring setbacks due to drought, soil impoverishment, or flooding, and increases in family size. Even published production figures such as those quoted in the Economic Survey for the year 1973 should not blind one to the facts.

Various kinds of millet, especially finger millet (Wimbi) and sorghum (Mtama), were once the staple food of the African farming population in Kenya, and this food was noted for its high nutritional value. Some kinds of millet continue to be used for brewing beer with a low alcoholic content, and this can be regarded as a valuable foodstuff. The discovery of the New World allowed three important crops to be introduced to East Africa namely maize, cassava, and sweet potato. Mixed with beans or at times with meat, maize has now become the staple diet almost everywhere in Kenya. Only in a few areas — western Kenya and the coastal region — has sorghum retained its equal or greater importance alongside maize in the feeding of the population. So far as home consumption is concerned, rice is cultivated by only a few farmers along rivers in western Kenya, and 90% of the country's needs are supplied from the Mwea Tebere Rice Irrigation Scheme. This rice is consumed mainly by the Asian group of the population, the majority of whom live in towns. Beans play an important role as sources of protein in almost all of Kenya's population groups. In some areas beans are capable of providing two or three crops a year. Cassava and sweet potato developed their important roles as reserve crops during periods of drought-induced crop failure as far back as the colonial era. Farmers in drought-prone areas are obliged by government decree to cultivate sweet potatoes and cassava. As the final crop in a rotation, cassava can remain in the ground for several years without perishing. Because of its modest demands on soil and attention, as well as its relatively high yields per acre, cassava a "lazy crop" is more popular among the farming population of the coast and western Kenya than is desirable for the community's health. Carbohydrates apart, cassava does not contain any important nutrients. Desirable quantities of animal protein, milk, and fat are only available to the better-off sections of the population, whereas the poorer population groups enjoy opportunities to consume these high-value and expensive foodstuffs only at irregular intervals. Limited supplies of fish are available on Lake Victoria and the coast, as noted in Chapter A. IV 6. The setting up of fishponds is only feasible in a few areas of Kenya and requires considerable "know-how". The development of a fishpond economy is for this reason still in its beginnings.

The food of infants after weaning is in particular very unbalanced and contains little protein; moreover, such feeding takes place too soon in many cases because of the onset of another pregnancy. It is known from certain studies that mothers usually breast-feed their children until the end of the first year and introduce supplementary food from the sixth month. This consists chiefly of maize gruel, and no vegetable or significant amount of protein is given. In rural areas about half of the mothers continue breast-feeding until the end of the 18th month; in urban areas it is rather less. Only 20% of the mothers continue with breast-feeding beyond the 18th month.

There are hardly any data on the clinical or subclinical malnutrition of children under fives, Blankhart (1974) being the only exception. Where 40% of the children under five are under-weight, the area is described as being of high risk, with shortage of food mainly contributing to this high prevalence of protein-calorie malnutrition. This was found, for example, in the Central Province at lower Mohito and Kituki, in the Eastern Province at Mbooni and Iveti-Kombu, in the inland coastal area near Kwale, and in the Eastern Province and northern Rift Valley.

According to the Annual Returns of Diseases of the Ministry of Health (Nairobi, 1969), nutritional disorders accounted for 6% of all deaths, being fourth in overall importance; of all cases admitted to government hospitals, this group accounted for 14.7% of the total and was third in importance.

The median treatment figures in the period 1965 to 1972 amounted to more than 25,000 annually in the case of kwashiorkor (code no. 286-6), to more than 6,000 for avitaminosis and other deficiency states (code no. 280-289), and to more than 14,000 cases for anaemias (code no. 290-293).

These statistics do not distinguish between kwashiorkor, marasmus and mixed forms of protein-calorie malnutrition.

In this considerable under-reporting must be noted, together with the fact that conditions of malnutrition are concealed behind the reported infectious diseases, particularly the lethal ones, and are responsible for the fatal results. The reporting of measles at a national level may serve here as an example, since it was the first statistical investigation of a relationship with kwashiorkor. The clinical progress of measles is influenced unfavourably by latent protein deficiency, and in the same manner a latent protein deficiency may become manifest through infection by measles (Morley, 1973).

From the aspect of geomedicine and epidemiology, this relationship can be very clearly demonstrated in Kenya (see Map 9 a).

The hospital recording rate of the 54 government hospitals confirm, after rank transformation, a relationship between measles and kwashiorkor for the years 1963 to 1972 (see Table 39). Making all due allowances for the weight and reliability of the data, this result can be fundamentally interpreted to the effect that, due to better

Table 39. Hospital Recording Rate (HRR) for kwashiorkor and measles of 54 Government Hospitals (1963 – 1972)

Kwashiorkor HRR (1963 – 1972)

	stanine rank (%)	lower 23% of hospitals with HRR <93	middle 54% of hospitals with HRR 93–465	upper 23% of hospitals with HRR >465	Σ
Measles, HRR (1963 – 1972)	upper 23% of hospitals with HRR >1133	0	4	8	12
	middle 54% of hospitals with HRR 221–1133	5	22	3	30
	lower 23% of hospitals with HRR ≦220	7	4	1	12
	Σ	12	30	12	54
	$\hat{\chi}^2 \gg \chi^2_{4,001}$			$\hat{\chi}^2 = 24,32$	

protein nutrition, the interaction effect of nutrition and infection is less evident in areas with a low kwashiorkor and measles recording rate than in those areas with a high kwashiorkor and measles recording rate. In spite of the inadequacy of the data, this analysis permits conclusions about a regional distribution of the nutrition problem in Kenya.

It is possible to recognise three different areal groupings:

(1) Areas with hospitals having a *low kwashiorkor frequency,* with a *low or medium frequency of measles* (12 hospitals):
   (a) Malindi, Mombasa, Galole.
   (b) Narok, Kajiado, Moyale, Makindu.
   (c) Nanyuki, Kapenguria, Kapkabet, Kisii.
This group includes coastal places with fishing (a), and areas where the population exercises a pastoral economy (b) and regions in which cattle keeping occupies a dominant position alongside arable farming (c). Although unable to keep many cattle because of the scarcity of farming land and the smallness of the farm units, the farming population of the Kisii Highlands carries on intensive agriculture which assures their food supplies.

(2) Areas with hospitals having a *medium frequency of measles and kwashiorkor* (26 hospitals):
   Bungoma, Kakamega, Kisumu
   Kitale, Tambach, Kabarnet, Maralal
   Kapsabet, Nandi
   Thomson's Falls, Nakuru
   Nyeri, Muranga, Kiambu, Tigoni, Nairobi
   Machakos, Kangundo, Kitui
   Voi, Taveta
   Kwale, Msambweni, Kilifi, Kipini and Lamu.
Considering the great variety of the areas listed above, it is not possible to give an explanation which would have general validity for this group.

(3) Areas with hospitals having a *high frequency of kwashiorkor, together with a medium or high measles frequency* (14 hospitals)
   (a) Thika, Kerugoya, Embu, Meru, Wesu
   (b) Kericho, Londiani, Molo, Naivasha
   (c) Isiolo, Marsabit, Lodwar, Garissa, Wajir
This category characteristically includes areas of high population density and land scarcity (a), as well as large-farm and plantation areas with a large proportion of agricultural labourers and their families (b). It is difficult to explain the high frequency of kwashiorkor in area (c), in which the population carries on a pastoral economy. Malnutrition is to be expected, insofar as population has become too numerous for the numbers of cattle, and the areas affected are often afflicted by drought; but bearing in mind that the diet always consists in part of milk or meat, the incidence of kwashiorkor ought not to be too high. Although this analysis of hospital statistics allows no direct conclusions concerning the actual conditions of nutrition, they must nevertheless be regarded as geomed-ical-epidemiological indicators despite the limited evidence they offer.

Apart from the Nutrition Survey of 1964 – 1968, it is in Blankhart's work (1974) that attention is drawn to the serious problem. Above all this author has called for further investigations of the relationship between nutrition and infectious diseases in children under the age of three; for investigations into simple, cheap infant feeding formulas based on locally and seasonally available and acceptable food to be prepared at home; for speedy and frequently repeated examinations of the frequency of slight protein-calorie deficiency symptoms in relation to weaning patterns, particularly in areas which have fo far remained uninvestigated, as well as over local habits of child feeding with a view to encouraging good habits and changing harmful habits, and an analysis of the existing practices of nutrition education. Despite all these proposals nutrition does not yet appear to play the role in the country's development policy which this problem now deserves and will deserve even more in the future.

Although the development plans for 1970 – 1974 and 1974 – 1978 contained references to a planned National Nutrition Council, WHO and UNICEF have carried out investigations into nutrition and the annual reports of the Ministry of Health occasionally mention local nutrition problems, no greater importance has to date been attached to this decisive factor in health and development.

A nutrition unit at the Ministry of Health supervises 100 nutrition field workers at the provincial and district level. In district hospitals, in mother and child clinics, and in mobile health teams, explanatory talks on nutrition, cooking instructions and demonstrations are carried out on the basis of foodstuffs locally available. Considering the size and distribution of the groups addressed and aimed at, this number is totally inadequate given the nature of the problem.

2. Anaemias

Anaemia is a major cause of ill-health among large groups of population in many countries, particularly those of the tropics. The most vulnerable groups are females in the reproductive age and children after weaning. A major contributory factor is the infestation with certain parasites, resulting in blood-loss through the intestines or bladder, as in the case of hookworm or schistosomiasis. Inadequate iron intake due to iron deficiency of nutrients or impeded iron resorption due to gastro-intestinal disorders of various kinds or other metabolic disturbances result in a negative iron balance, enhanced by blood-loss during menstruation or delivery in the case of the female, or high iron demand or protein-calorie malnutrition during infancy and childhood.

Although hospital statistics barely distinguish between the different forms of anaemia, iron deficiency plays the dominant role. It is followed by megalocytic anaemia as a result of folic acid and/or vitamin B 12 deficiency, anaemia due to malaria or sickle cell anaemia, and kwashiorkor (Young, Kondi and Foy, 1974).

The lack of capacity to work in the individual or population as a whole is difficult to measure and remains unknown in a country like Kenya, although it must be assumed to be considerable. Serious anaemias are a daily feature in medical practice, although there are distinctive regional variations. The actual distribution throughout the country remains unknown. The figures laid down in the hospital statistics provide, if anything, only a rough guide to the regional distribution of the problem.

Without entering into the particulars of the quantitative and qualitative problems of the relationship between hookworm infestation and anaemia, which has been investigated in Kenya by Foy and Kondi (1960), it should be noted that anaemias are quite frequently observed in the hospitals of the coastal towns where hookworm infections are also very commonly reported (see Map 8 and Chapter IV).

In the remaining parts of Kenya as well, a certain relationship can be recognised between the frequencies of anaemia and hookworm, and also between anaemia and kwashiorkor.

Another group, particularly typical for Africa, is that of the haemoglobinopathies. Foy and Kendall (1974) presented a survey which showed considerable variation in the geographical distribution of sickle-cell anaemia and the sickle-cell trait.

The highest frequencies were observed among the Luo in western Kenya, on the coast south of Mombasa, and in Taveta. Although the cause has not been completely established it is assumed that carriers of the sickle-cell trait experience a lower malaria morbidity and mortality than the carriers of normal haemoglobin, the former being subject to a positive selection as far as malaria is concerned. Other, rarer, haemoglobinopathies were also found sporadically, but no details of their frequency and geographical distribution are available.

VII. Diseases of Special Geomedical Relevance

Apart from the diseases described so far, the geomedical relevance of which is more or less evident from the ecological conditions of the disease vectors or the animal reservoir of the causal agents, or from special conditions of life or hygiene (i.e., the ecology of man), there are some other pathological conditions which are geomedically relevant in the narrower sense of the word. Endemic goitre or fluorosis are cases in point.

For years the observation of distinct regional variations in the occurrence of certain types of cancer in Kenya has provided particular stimulation for the undertaking of further investigations into cancer causation. Possible genetic or environmental factors which may contribute to this situation can be readily studied in the comparatively stable—although in life styles differing—population of Kenya.

1. The Geographical Distribution of Endemic Goitre

Hanegraaf and McGill (1970), the first to provide an extensive review of the relevant literature, arrived at the conclusion that the distribution pattern corresponds to that of other continents insofar as the inhabitants of mountains suffer more often from goitre than the remainder of the population. Endemic goitre appears to occur with marked frequency in the highlands of the central Rift Valley, the Aberdare Ranges, and in the Western Highlands. A detailed investigation was carried out by J. A. Munoz of the WHO Nutrition Unit during the period 1960 – 64. Goitre rates varying from 15 – 72% were discovered, the highest of these occurring in the highlands of the Rift Valley, in the Central Province, and in the Western Province (Bohdale et al., 1968). Carried out among school children, these investigations are not fully representative insofar as visible indications of a malfunctioning of the thyroid gland generally become apparent only in later years. Hanegraaf et al. (1970, 1971, 1974) therefore checked three rural areas near Naivasha, Kericho, and Machakos by conducting sample surveys among all age groups and examining iodine excretion, the protein-bound iodine in the serum, and the capacity for iodine uptake. There was a good consistency so far as epidemiologically established goitre, reduced iodine excretion, and increased iodine uptake were concerned. The authors concluded that in goitre areas where goitre occurs in younger age groups, the female sex, with the exception of the age group up to 6 years, is more prone to the disease, that cretinism does not appear, and that toxic disturbances of the thyroid gland are rare.

In the hospital recording rates the following hospitals stand out above all the others: Kericho (75 cases per 100,000 inhabitants in the catchment area as a long-term average), Nanyuki (46), Nyeri (33), Molo (17), Voi (16), Embu (12), Thomson's Falls (11), Nakuru (11), and Marsabit (10).

In 1970 preventive measures were started on an experimental basis by issuing iodized salt. Results have not been clearly established yet.

2. Fluorosis
(Mottled enamel)

While a reduced fluorine content in drinking water (< 1 p.p.m.) leads to susceptibility to dental caries, fluorine content of more than 4 p.p.m. among young people results in hypo-calcification of the dental enamel and thus to mottled enamel. Changes in teeth in accordance with this were observed mainly among schoolchildren in the Rift Valley (Fendall and Grounds, 1965a). As a result of volcanism, the surface waters contain excessive fluorine content: Lake Rudolf (13.1 p.p.m.),

Lake Naivasha (30 p.p.m.), Lake Nakuru (2.8 p.p.m.). Well water also contains an increased fluorine content (Bakshi, 1974).

3. Cancer and Other Malignant Tumours

Whereas the total incidence of cancer in Africa, especially, in East Africa and Kenya, may be lower than in Europe and America, there are certainly some tumours which are far more common in Kenyans than in Europeans or Asians or other non-African communities. On the other hand, those cancers are few which are characteristic of industrialized western countries, or of populations with a high life expectancy or a child and infant population no more at risk to early infectious diseases and death. The quite distinct differences of cancer occurrences within African communities or localities lead to the view that environmental, dietary, or traditional factors influence the causation of particular tumours. The first person to draw attention to this fact was Denis Burkitt, who described what later came to be known as *Burkitt's lymphoma.* In 1964 – 1967, he carried out a large cancer survey for the most frequent and conspicuous tumours—such as oesophagus, stomach, liver, and penis cancer, scar epithelioma, Kaposi sarcoma, and Burkitt's lymphoma (Burkitt et al., 1969, Cook and Burkitt, 1970). Since 1967, a national cancer register has existed in collaboration with the International Agency on Cancer and the United Kingdom's Medical Research Council (Linsell, 1974). Data from these surveys and this register are quoted here.

Cancer of the oesophagus shows a great variation among Africans; with few exceptions it is almost unknown in Europe. Cancer of the oesophagus is the most frequent cancer in the Kisumu area and throughout the highlands of the eastern Rift Valley, while in other areas it is almost unknown. It seems to be predominantly confined to the Luo and Luhya. No plausible ecological explanation has so far been found. Stomach cancer seems, on the other hand, to be rare.

Cancer of the penis occurs almost exclusively among those tribes, the Luo, Samua, and Turkana, who do not practice circumcision, regardless of the type of circumcision or the age at which it is performed. As *cancer of the cervix uteri* is the most frequent female cancer in Kenya, it shows an association with the pattern of distribution of penis cancer.

Cancer of the liver: Heptocellular cancer is known to be common throughout tropical Africa. In Mozambique it accounts for over 50% of all cancer, in Ethiopia for more than 30%. In Kenya it appears to be common everywhere. It may possible be associated with chronic liver damage, with malnutrition, alcohol, virus infections, or the hepatotoxic aflatoxin produced by yeasts on stored cereals.

The *kaposi sarcoma,* a tumour of the vessel-forming mesenchym, which leads to knot formations on the skin, is much more widely spread in Africa than in Europe; it occurs in Kenya, but even more so in Uganda.

Nasopharyngeal cancer is especially frequent among the Kipsigis and Nandi, as well as among other highland dwellers, whereas it remains practically unknown in the lowlands. It has been suggested that this is to be attributed to the climatically-conditioned protracted sojourns in the smokey and badly ventilated huts; these with their continuously smouldering acacia wood fires lead to chronic conditions of irritation and carcinogenic effects (Fendall and Grounds, 1965 a).

Among the *tumours of the infant age group,* it is particularly *Burkitt's lymphoma* which shows a distinct correlation with climatic zones in which malaria is transmitted throughout the year. There seems to be a relationship with the *Epstein Barr virus,* possibly in connection with malaria infections.

These few examples show that the study of the geographical distribution of tumours and the conditions of life of their carriers can make an essential contribution to investigations of cancer causation.

4. Diabetes

Among Africans diabetes is by no means as rare as had previously been assumed. The problem of diabetes in Africa is mainly characterized by the lack of medical facilities and at present the virtually non-existent possibility to organise proper management of diabetes by the patient and the medical personnel. Diabetes clinics exist only at the Kenyatta National Hospital in Nairobi and at the Coast Provinces General Hospital at Mombasa. Chronic diabetics will soon die from the consequences of the disease, although disturbances of the metabolism, so frequent in Europe, are rarer as a result of a diet relatively devoid of fats. The earliest clinical-epidemiological analyses of cases of diabetes in Kenya were undertaken by Darragh, Hutchinson, and Mngola (1971) from Nairobi, and Mngola (1974) from Mombasa. The analysis of the past ten years of the hospital recording rate for diabetes shows a remarkable frequency in the hospitals on the coast, in the Central Province, and at Kericho. A detailed investigation of the frequency of this distribution would be informative. The 12 hospitals with the greatest occurrence of diabetes are situated on the coast of the Indian Ocean: Lamu records an annual mean of 51 cases per 100,000 inhabitants in its catchment area, followed by Kipini (35), Malindi (27), Kilifi (13), Mombasa (21), Msambweni (25), and Voi (20). This observation confirms the often discussed frequency of diabetes among Semitic and Hamitic population groups. In the Central Province, Meru reports 45, Embu 35, Nyeri 30 and Nairobi 14; Kericho with 49 cases is the only hospital outside both these regions which has a higher number of cases. The remaining 77% of the hospitals for which registration figures are available report insignificant occurrence or non-occurrence of diabetes. These regional differences certainly have reasons which would be interesting to investigate.

D. Economic and Socio-geographical Classification and the Patterns of Health Services and of Disease Occurrence

Pre-industrial societies require closer adaptation to the ecological conditions of their region than do industrial societies; the latter are able to overcome naturally unfavourable factors by technical means and better "know-how". Nevertheless, ecological conditions provide only the framework for a society's economic activities. To which degree this framework is filled-in depends upon socio-economic circumstances and their historical origins.

The natural structure of the Kenyan territory, and the ecological conditions resulting from it, continue to emphasize these inter-relationships. In addition, a geomedical synopsis illustrates inter-relationships between factors of economic and social geography, geomedicine, and the pattern of disease distribution resulting from the conditions of life—even if for methodological reasons it is not always possible to quantify them.

This manner of approach permits a division into 15 more or less differentiated economic and socio-geographical regions, which will be described subsequently, including the conditions of health services and disease patterns previously explained in Chapters B and C. Special emphasis will be placed on the description of potential developments and their inherent health problems.

1. African Peasant and Pastoral Societies

a) Peasant societies in areas of rainfed agriculture over 1,500 m above sea-level: In rough approximation to reality, the 1,500 m contour should be regarded more as a broad band than as a line, above which the regions or core areas of economic development are to be found.

The mean annual temperature maxima remain below 26 °C.; the mean annual minima even fall below 14 °C.; at higher altitudes, they fall clearly below 14 °C. even though the highlands are situated on both sides of the equator (cf. Map 3).

In most years the quantity and distribution of the precipitation is sufficient for two crops, one during the main growth period of the long rainy season, the other one during the "short rains", in which there is a limit to the variety of crop species which can be profitably cultivated. But even during the rainy season, precipitation never reaches a level of intensity such as to impede economic activity very markedly. In particular the occurrence and distribution of animal vectors or the intermediate hosts of epidemic tropical diseases is already greatly reduced on account of climatic conditions; consequently they naturally no longer have the importance that they exercise in lower-lying areas. In particular the intensity of malaria transmission declines from this altitude upwards. Apart from a few exceptions on the western slope of the Rift Valley Plateau around the upper reaches of the Nzoia and its branches, or in the Nandi-Kipsigis area, malaria is practically speaking of no importance above 2,000 m.

Tsetse flies (Glossina spp.), which play an important role in the transmission of sleeping sickness in the lower parts of western Kenya and southern Nyanza, no longer occur; the same is true of phlebotomes, which are responsible for transmitting Kala Azar, the upper limit for which is found at about the 700 m contour. So too with water temperatures—these fall so much at this altitude that the snail hosts of *S. mansoni* and, above all, of *S. haematobium* cease to occur.

However, as far as other infectious diseases are concerned, the cooler climatic zones, in conjunction with hygienically inadequate dwellings and a high residential density, favour their distribution. A striking climatic inter-relationship can also be established in respect of intestinal helminths, the eggs and larvae of which infect man through the soil. Whereas the eggs of *Ascaris lumbricoides* (roundworm) show a marked resistance to temperature and dryness or wetness alike, the biotope of hookworm larvae is dependent on a much narrower temperature-humidity-milieu range. As a result, the relative proportion of hookworm infection in line with increasing altitude (i.e. decreasing temperature and humidity) becomes distinctly smaller when compared with roundworm infestation.

In the tropics, the quantity and distribution of precipitation is more instrumental in determining soil fertility than is the parent rock from which the soil has derived. In the Kenyan Highlands soils deriving from Tertiary or Quaternary volcanic masses predominate. As a rule these soils are deeply weathered and fertile; their fertility decreases with altitude, where the increasing precipitation intensity has leached and soured them at the upper limit of cultivation.

In addition the soils of the basement complex, as found in the Kisii Highlands or the Kakamega Highlands for example, allow intensive smallholder agriculture.

Already in pre-colonial times the wooded highland districts of Kenya made up the settlement areas preferred by cultivators, while the vast, grassy plains were dominated by Masai herdsmen. Since the colonial period the concentration of population in the highland areas has progressed at an even greater speed; it was encouraged during the colonial period by the policy of the administration, which permitted land purchase by Africans only in the so-called reserves. In the post-colonial period emigration from these densely settled high altitudes has taken place to only a relatively limited extent. This will be discussed in greater detail below.

At heights above 1,500 m in Kenya, the population density almost everywhere reaches more than 100 persons to the square kilometre. The predominant density is, however, between 200 and 500 inhabitants per square kilometre. In some places such as the Kikuyu settlements north and north west of Nairobi this density is surpassed.

Increase in population and the increased demands for cultivable land generated thereby have led to the destruction of the original forests and the expansion of settlement on the Aberdare Range and Mt. Kenya up to the altitudinal limit of cultivation. So too, steep slopes were brought under cultivation in many places with the result that considerable damage due to soil erosion has occurred. The scarcity of land resulting from population increase has led to changes in land tenure, which have been described in section A IV 1 cc.

Maize forms the staple food of the peasant population living in the highland areas above 1,500 m, supplemented in many cases by legumes. Finger millet, which was once the staple food of the population in the highlands, is now only grown in small quantities. The diet is frequently supplemented by bananas, various sorts of vegetables, sweet potatoes and other plants.

The ecological conditions in the Kenyan highland areas above 1,500 m favour the cultivation of valuable cash and export crops such as arabica coffee, tea, pyrethrum, various kinds of temperate zone vegetables and fruit, potatoes, wheat, and much else.

The ecological conditions also facilitate the modernisation of husbandry. In the Kenyan highlands high-yielding European dairy cattle can be kept, so that the farmers are in a position to derive income from milk sales. Animal husbandry must, however, be integrated with farm unit generally. Since cattle can no longer be put to pasture in the same way they were previously turned out in the bush, cultivation of fodder and the preparation of seeded pasture have become the preconditions for dairy farming. Increasing scarcity of agricultural land has led to a substantial decline in traditional animal husbandry, and the effect has been to make the supply of protein to the population problematic.

Hoe cultivation still remains the dominant agricultural technique and seems likely to continue to do so, given the steepness of the terrain and the small size of most farms which militate against the keeping of draught oxen. Even the traditional hoe cultivation itself calls to mind forms of horticulture rather than agriculture. Intensification of peasant agriculture might well lead more farms to adopt the character of market gardens.

The agrarian reform depicted in Chapter A. IV. 1, a) has progressed furthest in the Kenyan Highlands, and the process of individualisation of property in land may be regarded as completed over large areas.

However, commercialisation of agriculture has prepared the way for increasing differentiation among what was once a more egalitarian peasant society. Successful farmers can enlarge their holdings by acquiring further acreage and increase their production for the market. Often these successful farmers are not farmers in the European sense of the word, for they leave their wives or paid labour to run the farm, while they follow some other calling in the administration or business. Gavelkind inheritance has, on the other hand, fragmented very many farms to such a degree that their heirs are unable to win a living from the land. Although the division of a farm is not effective without the permission of the Land Registry, that has not prevented ready buyers from dividing such farms. The sub-division is not entered in the Land Register, and this uncertainty over the law is given added impetus and impedes the development of peasant agriculture, as Fliedner (1965, p. 86) has shown for the Kiambu area.

An essential aim of the policy of peasant development is the improvement of nutrition among the peasant population, but the aim is only realised in part. Many of those who produce high-quality foodstuffs such as vegetables, milk, and meat, economise in the feeding of their own families in favour of acquiring cash through sales. More detailed investigations of this aspect of development are so far lacking.

Equally unknown is the extent of unemployment and underemployment on the excessively small farms of the Kenya Highlands.

On the other hand, appropriate agricultural land utilisation can be observed in the Kenya Highlands. When visiting progressively-run farms, it is difficult not to conclude that the productivity of peasant agriculture could be raised significantly, as well as the nutritional standards of the peasantry. These areas above 1,500 m have the best infrastructure of health services, especially the non-governmental health services (see Chapter B. 5).

b) Peasant societies in areas of rainfed agriculture between 1,000 and 1,500 m above sea-level: The peasant farming societies live in ecologically very differently endowed regions, with the one common feature, however, that high value cash crops such as arabica coffee, tea, pyrethrum, potatoes, various types of European fruit and vegetables do not flourish for climatic reasons. In these regions are the extensive plains of western Kenya, situated at between 1,200 and 1,500 m above sea-level; their soils derive from granitic material. In the 8 – 11 humid months there is a long-term annual mean rainfall of 1,000 to 2,000 mm, and as a rule two crops per year are harvested.

The problem areas are the immediate coastal strip on Lake Victoria—due to the small amount of precipitation—and the Kano Plains in the region of the Kavirondo Gulf, where the heavy, clayey black cotton soils are very hard to till with the agricultural implements available to the farmers. During the dry season the soils are rock hard, whereas during the rainy season large tracts of land are inundated. Nevertheless, the area is densely settled, but agricultural development remains in its infancy.

Similarly problematic are the agricultural conditions and the nutritional status of the populations in the Machakos and Kitui Districts. The soils of the basement complex in the lower hill country are less productive, and in 5 to 6 humid months there are about 700 to 1,200 mm of precipitation, varying according to the altitude of the area concerned. In both these districts drought-conditioned famines occur frequently (see Maps 3 a and 3 b).

Overpopulation in the Machakos District had already reached serious levels in colonial times, and it led to destruction of the forest and soil erosion. The outcome is to be seen today in the run-down impression Machakos District creates. Kitui District was much less densely populated. At the present time one can observe how the cultivated area is being extended. In the areas of rainfall cultivation below 1,500 m the population density is far less than that above 1,500 m, and only in a few locations in the north and west of the Kagamega District does the density rise to 300 persons per square kilometre. In other regions of Kenya, population densities between 100 and 200 inhabitants per square kilometre predominate. In the Machakos and Kitui Districts below 1,500 m it never rises above the 200 persons per square kilometre mark. Those areas of western Kenya which receive abundant rainfall could still tolerate an increase in population density, but in the remaining areas the agricultural carrying capacity is being exceeded, as has already been shown, due to the low agricultural potential and the present farming techniques.

Apart from maize, the main food crops cultivated in western Kenya are sorghum, legumes, other millet varieties, cassava, and sweet potato. Groundnuts, sesame, sugar cane, and rice (which is grown on river flats in some places) are of lesser importance in the nutrition of the population. In the Kitui and Machakos Districts maize is supplemented by legumes as the staple diet. Sorghum and other drought-resistant varieties of millet complete the picture. Cotton and in the Kitui and Machakos Districts sisal and castor oil are the only crops grown for the market. Surplus production of subsistence crops is also sold following good yields, but in poor years farmers are only able to obtain a little cash from sales of hedge sisal or castor oil fruit. Severe poverty among the peasant population and crop failure due to drought are specially marked in the Kitui and Machakos Districts at altitudes below 1,500 m. In the Kisumu District on Lake Victoria, farmers living in the vicinity of large sugar cane plantations are now able to cultivate sugar cane and sell it to the processing plants of the plantations.

As a result of the longer fallow periods needed by the soils to allow regeneration of their fertility and because of lower population density in most areas between 1,000 and 1,500 m. above sea-level, more extensive areas are covered with bush than at altitudes above 1,500 m. This enables the peasants to keep more livestock. In the marginal cultivated areas around Machakos and Kitui, livestock is essential as a supplementary source of food, which thereby safeguards the existence of the peasant families. Highly productive European dairy cattle cannot be kept, however, the keeping of livestock is not integrated with other farming activities, as described in Chapter A, IV, 1, a.

In western Kenya, however, and in the Luo tribal area around Lake Victoria, draught oxen are kept and ploughing has largely superseded hoeing. It is customary to practice semi-permanent cultivation, with the number of years of cultivation exceeding those of fallowing.

While the process of Land Adjudication and the Registration Programme may be regarded as completed in the densely settled altitudes above 1,500 m, it is still in progress in the areas of rainfed cultivation below 1,500 m, where in many places farms are still awaiting demarcation, survey, and entry into the Land Register. Although the chances of commercialisation of agriculture are less than in the climatically favoured high altitudes, efforts are being made to encourage the cultivation of cash crops. In this context the cultivation of maize assumes special significance as it is this crop that Kenya has begun to stockpile in order to be able to provide for the population from its own reserves in the event of crop failure.

By contrast with the agriculturally active areas above 1,500 m, the altitudes 1,000 – 1,500 m are characterised by a number of infectious diseases, the transmitters or intermediate hosts of which are favoured by climatic conditions.

In these altitudes malaria is transmitted almost throughout the year. In western Kenya and Nyanza, as well as in the Machakos-Kitui area, the number of infections is accordingly high. Even if no measurements of the high morbidity rate in these areas were made of the retarding influence on the development of agriculture, it must still be assumed that malaria constitutes an unfavourable factor not altogether unimportant.

In southern Nyanza in the Lambway Valley, there is a last focus of sleeping sickness. Other foci in the Siaya-Kisumu District were brought under control only in the course of the last decade. But then as now the entire lakeshore area of Lake Victoria, and inland, is to be regarded as a potential area of tsetse infestation and thus at risk to sleeping sickness, even if the present control programmes realize the promise of long-term success. In South Nyanza and in this region of the Siaya— Kisumu District there is a reciprocal effect of agricultural activity, the style of land utilisation and the ground-water level on the one hand, and the tsetse fly distribution and development of sleeping sickness outbreaks on the other; like malaria and schistosomiasis, there is a clear, positive correlation between agricultural development measures and the control of disease vectors. At these altitudes both these areas, West Kenya-Nyanza as well as Machakos-Kitui, are severely infested by schistosomiasis. The distribution of the snail vectors of *S. mansoni* and *S. haematobium* is particularly favoured by the prevailing ecological conditions in these zones, but it is also a result of population pressure and planned as well as completed irrigation measures for agriculture on a small, individual as well as large, cooperative scale, which are required to make up for the lack of precipitation (see Chapter C IV, 1, D 1, e and D 1, f).

Hookworm plays a special role too due to climatic conditions (see Chapter C IV, 1).

For unknown reasons even leprosy appears to be more widely spread in these zones than elsewhere (see Chapter C II, 2).

In respect of medical facilities, these regions are numbered among the relatively favoured ones. There is a good road network, population density is high—with the result that spatial effectiveness is of a high order. Historical events account for the positively good provision of health services offered by the churches.

c) African and Afro-Arabic peasant societies in areas of rainfed agriculture in the coastal lowland: The habitat of the African and Afro-Arabic peasant societies in the coastal lowlands is separated from the Kenyan core area by the poor, arid areas of the Nyika. In historical times the population of this region had few contacts with the interior of the country, for not only the inhospitability of the Nyika itself but the warlike attacks of the Masai and Galla made caravan traffic difficult. Nevertheless, the coastal population maintained cultural and economic links with the peoples of the Arab and Indian culture areas. Numerous ruins and historic finds testify to the Arab dominance of the coast, and for 200 years,— that is from 1492 to 1692— the Portuguese maintained a base at Mombasa for the sea-route around Africa to India, which they then once more lost to the Arabs. Even today there is only the one major link from the coast to the highlands, namely that from Mombasa to Nairobi. This major connection consists of an extraordinarily busy main road, the railway already mentioned, and the air route.

Morgan (1973, pp. 173 – 190) divides the Kenyan coast into a northern, sparsely settled section and a southern, relatively well-populated one, the economic centre of which is Mombasa, with good reason. But the agriculturally usable area, with in the main 6 to 8 humid months, is very narrow and only extends about 30 to 60 km from the coast into the interior. At this point the aridity limit for agriculture is to be found and linked with it the settlement margin for the peasant population. The ecological potential of this region is not yet fully utilised. For the development of the area a further build-up in population density would be desirable. In most places the population density remains below 100 persons to the square kilometre, only rising to values of up to 400 persons per square kilometre in the northern catchment area of Mombasa. These people are not families engaged solely in farming, however; many Africans from this district commute to Mombasa to work.

The form of agricultural land use differs from that of all the others in Kenya on account of the cultivation of fruit trees, in which the coconut palm is the most characteristic, although mangoes, citrus, and cashew trees are also widespread. The population's staple foodstuff is maize, and in the drier areas, cassava; cash crops include not only the tree crops mentioned above but also cotton.

Animal husbandry, and in particular the keeping of cattle, is made difficult by ticks and tsetse flies.

Shifting cultivation including burning the cleared bush still appears to be the most widely practiced form of agriculture. Cultivation is also carried on beneath the tree crops. The strong Arab influence finds its expression in the land tenure: as early as the colonial period most agriculturally useful land was recognised as private property at a time when property titles in the European sense did not exist in the other African peasant societies. The size and structure of holdings also differ from those of the purely African peasant areas. In some places the peasant holdings are interspersed with large farming units, among which the sugar, sisal, and coconut plantations are of special economic importance. Many members of the farming population have found additional employment through tourism (cf. Chapter A V, 2).

Besides malaria, which is particularly common in this hot and humid coastal strip, a special role is played by the *Wuchereria bancrofti* filariasis (though at first sight not much in evidence), and by *S. haematobium* schistosomiasis. So too infectious hepatitis appears remarkably frequently in the Annual Returns of Diseases of the hospitals. Hookworm infestation predominates over that of roundworm and tapeworm (see Chapter A IV, 1). The settlement schemes and irrigation schemes established at Taveta, Voi, and Galole (Hola) in the coastal hinterland have promoted the occurrence of these diseases. By comparison with the northern part of the coastal area, leprosy predominates in the southern half—a fact attributed by some authors to the immigration or settlement of population groups particularly susceptible to leprosy infection.

The provision of medical services is particularly effective in the coastal strip. Here, too, the concentration of the population along a narrow strip with good transport links along the coast have combined to increase it. In the face of this, the sparse population of the coastal hinterland suffers from similar disadvantages to those people living along the Tana River; exceptions to this are the townships of Galole, Voi, Taita, Taveta, and Wesu.

d) African peasant societies in marginal areas of rainfed agriculture: The marginal areas of rainfed cultivation occupy an extensive transitional area between the Kenya Highlands and the broad, arid strip which divides the coastal region from the highlands. But on the Kavirondo Gulf and in the Cherangani and Ilikamasia Hills there are also relatively narrow belts of land in which the ecological conditions prove marginal for agricultural activity. However, these belts are so narrow that it was not possible to consider them in a classification on the scale of 1 : 2 million.

In the marginal areas of rainfall agriculture, located in the Machakos, Kitui, Meru, and Embu districts (inhabited by the Kamba and in the north by the Tharaka), the population density remains very low. In some areas there are less than 40, and in other areas even less than

20, inhabitants per square kilometre. The long-term precipitation average is only 500 – 700 mm. As is evident from Map 3, the regularity with which these rains fall is poor. The soils of the basement complex are soon exhausted and require prolonged periods of fallow in which to regenerate their fertility. Accordingly the land needs of the population are high. The most widely practiced form of cultivation is the shifting type with bush burning; drought-resistant crops such as the various species of millet, cassava, legumes and, in the somewhat more favourable areas, even maize are grown. The crops are very often at risk when the rains fail to fall in sufficient quantity. To safeguard their existence, the population is obliged to keep adequate numbers of cattle, and cash is earned mainly through the sale of these animals.

In these areas, malaria has a meso-endemic prevalence. The region between 500 – 700 m altitude is known as a natural focus of visceral leishmaniasis (kala azar) with low endemicity in the northern part of the Rift Valley, the Kericho Valley, in Sambu District, and in West Pokot (see Map 7). The main area of kala azar in epidemic or endemic form is found in the deeper parts of Meru and Kitui District and along the Tana River (see Chapter C I, 3 a). These areas with low population density have a marginal infrastructure of health services and therefore the knowledge of the prevalence of diseases is relatively limited.

e) African peasant settlement schemes in areas of rainfed agriculture: The colonial administration had already aimed at alleviating the land shortage in the densely populated highlands above 1,500 m. In fact it had carried out several settlement schemes in uninhabited areas suitable for rainfed cultivation, as for instance the Kibirichia and Kirua settlement schemes in the Meru District. The settlement schemes outside the Large Farm Areas were carried out by the African Land Development Board.

The present government has continued with this policy by transferring peasants with little or no land from densely settled areas to less densely settled ones. In particular, several settlement schemes are to be carried out in the coastal area. Already the Shimba Hills Settlement Scheme has provided holdings and homes for 2,000 Kamba families. But the largest settlement programme carried out to date in Kenya was the so-called "One Million Acre Settlement Scheme", 1962 – 1966, in the course of which 400,000 ha of agricultural land was purchased from European farmers and redistributed among 35,000–40,000 African settler families. Common to all settlement schemes is the considerable control and influence exercised by the agricultural administration over the settlers and the cooperative organisation of the supplies of seeds, tools, fertilizers, etc., as well as the sale of produce (Hecklau, 1967, p. 141). The size of these settler holdings was not schematically decided but determined according to net income targets which the settler is intended to achieve. Sales of farm produce are intended to provide a certain net income for the owner over and above the payment of the half-yearly redemption instalments for the 30-year-

credit granted for the purchase of the settler's holding. In addition, every settler ought to be able to produce from his land all the food necessary for the supply of his own family. According to the level of these net income targets low and high density schemes are classified as such, with those with higher income targets and correspondingly larger farm units and lower population density, and those reaching a higher density of population in which the settlers received smaller holdings intended to produce smaller incomes. The distribution of land took place chiefly over the period 1963 – 1969, with small additions in the years following. The extent of land distributed up to the year 1972 is shown in Table 40.

Farm budget plans have been drawn up for the settlers taking into account the ecological conditions of the area concerned and the marketing conditions affecting the agricultural produce. These plans contain recommendations as to which crops the farmers should grow for the market. These are, according to ecological conditions, maize, tea, coffee, pyrethrum, sisal, passion fruit, sugar cane, and others. Almost all the settlers received European dairy cattle for dairy farming. For their own supplies, most settlers grow maize as a staple food.

When one considers the growth of the African population it is obvious that the settlement of some 35,000 peasant families on former large farm land cannot signifi-

Table 40. Land re-settlement. Area planned and plots allocated, 1963 – 1972 *

Province and Type of Scheme	No. of hectares planned	Plots allocated
Eastern		
High Density	13,345	770
Low Density	—	—
Total	13,345	770
Central		
High Density	153,312	13,315
Low Density	26,655	1,414
Total	179,967	14,729
Rift Valley		
High Density	103,538	6,385
Low Density	19,656	1,129
Total	123,194	7,514
Nyanza		
High Density	18,733	1,774
Low Density	17,695	1,938
Total	36,428	3,712
Western		
High Density	67,716	6,947
Low Density	8,542	505
Total	76,258	7,452
Total	429,192	34,177

Source: Ministry of Lands and Settlement.
* The figures included refer only of holdings allocated to individuals after the subdivisions of the former European owned farms. The system of subdividing these large estates has now stopped and new farms are being settled on a cooperative basis.
Source: Statistical Abstract 1975 Table 77 (a)

cantly lessen the demand for land in the areas of peasant agriculture.

Experience shows — according to the Development Plan for 1974 – 1978 — that most agricultural products can be grown very successfully on small farms. It is therefore planned to break up and share out still more large farms provide the settlers with an effective introduction to the modern monetary economy and practice in modern methods of land utilisation (Republic of Kenya: Development Plan 1974 – 1978, Part I, p. 199, Nairobi, 1974). "The politico-economic and also the agro-geographical significance of the reform of land ownership can be seen, however, in the resulting opportunity to by intensive counselling, training and supervision. Unlike other parts of Africa the Kenyan settlement areas thus present an opportunity to form an agricultural élite which could set the pattern for the entire country" (Hecklau, 1968, p. 263).

Depending upon the altitude and climate of the new settlement areas and in accordance with the pre-existing disease pattern of the resettled population, there are accordingly consequential health problems.

Although climatic conditions in respect of insect-borne diseases (especially malaria) can be described as favourable to man in settlements above 1,500 m and even the danger of schistosomiasis transmission is almost non-existent (because of the absence of snail vectors in natural waters and the non-existence of artificial irrigation installations), these two problems seriously affect such projects, as in the case of the Shimba Hills Settlement Scheme on the East Coast.

In principle there exist specific health risks of another kind, which are difficult to grasp in data terms and have not been investigated so far. The transplanting of groups of population, even if to ecologically more favourable regions, is accompanied by stresses, particularly of a psycho-social kind, which have until now received insufficient attention. Changes in the manner of settlement, nutrition, and economy require adaptation by man, even though they may superficially appear to improve the living conditions. A characteristic danger is to be found in the worsening of the nutritional status due to increased production for the marked and ill-advised expenditure on consumer and luxury goods or status symbols rather than on improvements in nutrition and of

hygiene standards. The generally inadequate medical services coinciding with the discontinuation of traditional, though in regard to their significance for health often under-rated, ties and systems of social security and health often results in an increase in infant and child mortality where a decrease ought to be anticipated among generally young families having high birth rates.

Resettlement of population implies on the one hand, potential carriers of diseases and of hosts of various parasites, and on the other, contact with autochthonous germs, insofar as disease reservoirs may have been present in the new settlement areas in the form of population groups already settled there.

Reciprocal exchange of germs and diseases may occur. In such cases it would seem that tuberculosis, poliomyelitis, and hepatitis have spread. In the Shimba Hills Settlement Scheme the Ministry of Health's Division of Vector-borne Diseases is successfully taking measures aimed at controlling malaria and schistosomiasis.

f) African peasant societies (tenant farmers) in state-organized irrigation schemes: It is estimated that around 200,000 ha of land in Kenya is suitable for irrigation. But owing to the high costs of development required to set up irrigation schemes only 8,471 ha of irrigated land had been developed up to 1975. The large irrigation programmes in Kenya are under the control of the National Irrigation Board.

aa) Mwea Tebere irrigation scheme: The most successful irrigation scheme at present is the Mwea Tebere Irrigation Scheme, comprising 5,379 ha and located on the southern extremities of Mt. Kenya at heights of about 1,200 m above sea-level. In 1971/72 the land was divided among 2,578 tenants, each cultivating four plots of about an acre apiece, a total of 1.8 ha of land (cf. Table 41).

The tenant farmers are required to follow a strict discipline of the annual sequence of tasks; failing that, they run the risk of having their lease terminated. Certain jobs are carried out with heavy machinery and by the state, others are dependent upon the tenant's own initiative. It is not intended to make the work fully mechanised, since it is more important to create a secure

Table 41. Irrigation schemes under the control of the National Irrigation Board, 1974/75

Irrigation Scheme	Hectares cropped	Number of Plot-Holders	Crop	Yield (M. tons)	Gross Value of Crop K£	Payments to Plot-holders
Mwea	5,379	2,917	Rice	28,423	1,349,328	902,729
Ahero	1,534	445	Rice	2,467	148,020	61,855
Tana River	856	585	Cotton	2,288	228,095	146,450
Perkerra	370	346	(Onions	1,267		
			Chillies)	188	118,000	58,944
Bunyala	332	112	Rice	674	30,351	9,559
All Schemes	8,471	4,405			1,282,154	1,873,794

Source: Statistical Abstract 1975, Table 96

existence for landless peasant families than to achieve the greatest output with the least input. The peasant families live in 19 small settlements, in which they have built their own dwellings. Schools were put up using communal labour.

In the Mwea Tebere Irrigation Scheme nine-tenths of Kenya's rice needs are met. It is to be enlarged to a size of about 6,000 ha, which would then include all the irrigable land. A detailed investigation of the Mwea Tebere irrigation scheme has been carried out by Golkowsky (1969).

bb) Ahero and Kano II irrigation schemes: The Ahero and Kano II irrigation schemes are situated in the Luo tribal area on Lake Victoria, where land is scarce and ecological conditions difficult. About 50,000 ha of land in the Luo area might be suitable for irrigation. The Kano Plains with their heavy, clayey alluvial soils particularly in need or regulation of their hydrological cycle. During the rainy season the soils are inundated, but if the rains do not set-in they dry out quickly and crops are at risk from drought. About 14,000 ha in the Kano Plains are assumed suitable for irrigation. So far, however, only the Ahero Irrigation Scheme is completed. In this scheme considerable difficulties are still to be overcome, which are brought about by attack from pests and low yields. So far rice has been the main crop, but experiments are being conducted with the cultivation of sugar cane, groundnuts and soya beans. The Kano II Irrigation Scheme is an experimental project which will extend to 1,280 ha when it is completed.

cc) Irrigation schemes on the Tana River: Substantial preliminary investigations have shown that 150,000 ha of land could be irrigated in the area of the Tana river. The greatest problem is that the potential irrigation terrain lies in an extremely sparsely settled and semi-arid region, far from Mombasa and the densely populated highlands of Kenya. Even today the poor road leading from Garissa via Galole to Malindi and Mombasa can become impassable in the vicinity of the Tana river during the rainy season.

The Galole Pilot Irrigation Scheme was started in 1957 and resulted in the cotton irrigation project near Hola. At first the yields from this scheme were low, but considerably increased outputs have been achieved since 1958. As cash crops maize, groundnuts, cow beans, and chick peas are grown for supply to the local population. The development of the irrigation scheme is to be supported by further public investments of K£ 369,000 over the planning period 1974 – 78. Almost half of this amount is to be used for the construction of a cotton ginning plant. (Republic of Kenya: Development Plan for the Period 1974 to 1978, Part I, p. 225, Nairobi 1974.)

Forty kilometres upstream from Hola another irrigation scheme is planned in Bura and this will comprise some 16,000 ha. Until 1978 an initial section of 4,000 ha is to be cultivated.

dd) Perkerra irrigation scheme: The Perkerra Irrigation Scheme was started in the sixties. Its situation is very remote, that is in the vicinity of Marigat south of Lake Baringo in the Rift Valley. Here the Njemp tribe practiced irrigation culture in pre-colonial times. However, the Perkerra river, which had irrigated the Njemps' fields, changed its course and the Njemps were evidently unable to continue irrigation culture in the rivers new surroundings. Even at the present, the modern irrigation scheme has to contend with considerable difficulties. Distance from markets, attacks from pests as well as human problems among the tenants have all helped to restrict the development. Cash crops grown there are onions and chillies.

ee) Irrigation schemes in the Yala Swamps area: A small irrigation scheme has already been completed near Bunyala, namely the Bunyala Irrigation Scheme. The Yola river which now runs into Lake Kanyaboli and the adjoining Yala Swamps is to be diverted by a canal. It is hoped that by lowering the water-level of the lake, about 200 ha of the swamps can be drained and then cultivated and irrigated using water from the canal. If the plan succeeds, part of the Yola Canal water will be led through the old river bed back to the lake, from whence it will hopefully distribute across the entire drained area of the former Yola Swamps through a system of canals. It will then be used for the irrigation of the newly-won fields (Hecklau 1970, p. 489). About 16,000 ha of land could be cultivated in this manner, and the development works are already in progress.

Particular health risks are common in such co-operative irrigation projects at altitudes below 1,500 m. Apart from what has been previously mentioned in Chapter D 1, d, there is a danger of artificially creating breeding places for the mosquito vectors of malaria and the snails which act as intermediate hosts of schistosomiasis The malaria situation is largely under control, partly because these areas are in any case holo-endemic and stable, and their new settlers already possess more or less distinct, acquired immunity or premunition. On the other hand suitable larvicidal and chemoprophylactic measures applied by the Ministry of Health's Division of Vector-Borne Diseases in the controlled settlement areas restrict the vectors of malaria and have reduced gradually the role of parasites among the population. The danger is essentially one of re-introduction from outside by new settlers, of the development of resistance of the mosquito vectors and malaria plasmodia towards the chemical substances applied, or of economic and organisational difficulties in keeping up the control measures which are demanding of personnel and finance.

A problem much more difficult to control is schistosomiasis (see Chapter C IV, 2). In every irrigation scheme, whether already in existence or still at the preparatory stage, there is the potential danger of intensive schistosomiasis transmission. In the area of the Mwea Tebere Irrigation Scheme, following massive snail infestation and infection of the population, the application of modern molluscicides at first succeeded in breaking the transmission cycle. The population received systematic treatment, with the result that the problem is now largely under control.

In the remaining projects no efforts have as yet been made in this respect. The snails are known to be intermediate hosts and the newly settled population is as infected by natural contact as is the autochthonous population. So far systematic control measures have not been introduced, although very important investigations have been carried out on the spot instead, with the result that the scientific basis for control at an appropriate time is at least available.

It is only in the Perkerra Irrigation Scheme that schistosomiasis does not occur, although the shores of nearby Lake Baringo are infested. Ecological conditions in the inner area of the irrigation scheme are not favourable to snails.

g) African peasant societies with irrigated cultivation: Irrigated cultivation existed in several regions of Kenya even in pre-colonial times—and that not solely among sedentary farmers, but also among herding families in the semi-arid areas of northern Kenya. Irrigated cultivation was practised on the slopes of the Cherangani Hills and on the Elgeyo Escarpment by several small population groups which are now considered to belong to the Pokot, the Elgeyo, and the Marakwet tribes. In this form of cultivation, rivers were skillfully diverted along the slopes and onto their fields. This simple irrigated cultivation leads to severe damage by soil erosion, and notwithstanding this irrigation, the fields must be shifted every few years when their carrying capacity is exhausted. The most important crops grown on irrigated land are sorghum, maize, sweet potato, finger millet, with lesser ones such as bananas, vegetables and groundnuts. Here and there cassava occurs as the last crop in a rotation. In some places the government has made attempts to substitute this traditional irrigation cultivation by modern irrigation schemes. But distance from markets and the thereby reduced opportunities for the sale of cash crops, vermin attack and the scant enthusiasm for progress among the farmers have greatly impeded modernisation of irrigated cultivation and especially the cultivation of cash crops. On the Elgeyo Escarpment chilli peppers and onions,—and in some places even a variety of vegetable—make up the most important cash crops.

Traditional irrigated cultivation, in some parts still conducted as in the days before the arrival of the Europeans and·in others more or less improved or even replaced by smaller irrigation schemes, may also be found in various parts of the country, as for instance among the Kamba in the Machakos District, among the Pokot in the middle course of the Tana River (which offers the largest irrigation potential in Kenya), and in the Taveta District on the Tanzanian border (where 6,000 ha are suitable for irrigation).

"A number of minor irrigation schemes have been established at various locations, mostly in the arid northern and eastern parts of the country where suitable water supplies exist. These include schemes in West Pokot and Turkana, at Garissa, Isiolo and Mandera, and several places on the lower Tana River. In total these schemes include more than 1,000 hectares of irrigated land. During the current Plan period, minor irrigation schemes will be developed on a much larger scale than previously, primarily as a means of improving food supplies and obviating the need for famine relief in the arid areas." (Development Plan 1974 – 1978, Part I,,p. 225.)

h)African pastoral societies in predominantly arid and semi-arid areas: Roughly estimated, the herding population of Kenya, most of whom are subsistence pastoralists, occupies three to four-fifths of the country. Due to ecological factors these regions are not agriculturally utilisable. Extensive regions receive less than about 500 mm of precipitation during less than three humid months. Only in some mountain massifs like Mt. Marsabit, Lloto Mountain, the Karabukot Hills, and the Masai settlement area in south west Kenya on the southern foothills of the escarpment does the amount of precipitation attain a level and favourable distribution such that agriculture is technically possible. However, the cultivated areas are so small that they cannot be mapped on a scale of 1 : 2 million. The crops only serve to supplement the extraordinarily critical nutritional position of the herding population, to be described later.

The herding population is almost exclusively composed of non-Bantu. The habitat of the Hamitic Somali, Boran, Rendille, Samburu, and Gala is found in the north east quarter of the contry, and extends as far as the south east. Between Lake Rudolf and the Uganda border are the Turkana, and in the south west of the country, in the Kajiado, and Narok Districts live the Masai, once the rulers of a vast territory in East Africa. In Kenya they were confined to these two districts following the treaties of 1904 and 1912, but their territory extends far into Tanzania.

Herdsmen are almost exclusively subsistence pastoralists; that is, they only rarely sell their cattle in order to obtain money for the purchase of food or the payment of taxes and other expenses. In all areas of pastoral economy, the population density is extremely low. It amounts to fewer that 10 persons per square kilometre, and over large tracts it even falls below 4 per square kilometre. The agricultural carrying capacity in areas of pastoral economy with less than 635 mm precipitation a year could amount to between 2 and 7 persons per square kilometre in Kenya, if modern pastoral methods were applied and the herding population fully integrated into the monetised economy (according to L. H. Brown: Development of the semi-arid areas of Kenya. Nairobi, 1963). This would require that the herdsmen sell the greater part of their stock and buy foodstuffs. At present such preconditions are only rather imperfectly represented in the areas of subsistence pastoralism. The situation is aggravated by the uneven distribution of the herding population across the arid areas, large bush-covered tracts being avoided because of lack of water and tsetse fly infestation. Moreover considering the fact that the carrying capacity of semi-arid areas remains extremely low under primitive subsistence pastoralism and un-rationalised management, it becomes evident that under present conditions many areas are grossly overpopulated. Given constant lifes styles and economy, the nutritional situation and thus the future of most of the herding tribes remain very problematic indeed. The growing number of cattle owned by the increasing human population leads to destruction of vegetation, which in turn causes soil erosion to such a degree that the carrying capacity of the herding population's habitat is diminished.

Constituting as it does the exclusive basis of life for the herding tribes, cattle ownership presents great differences in the numerical combination of the herds with cattle, sheep, goats, and donkeys in the areas of subsistence pastoralism. Cattle plagues which were once the cause of high mortality among the cattle have now become rarer with the availability of modern veterinary medicine. But very high losses among the cattle do still occur as a result of prolonged droughts. Under such circumstances the nutritional status of the pastoralists grows so critical that food supplies paid for out of public funds are the only means of preventing large-scale famines. Due to inadequate supplies of food, malnutrition can be observed in the population in many parts of the country.

Most of the herding tribes keep zebu cattle, goats, and sheep. In northern Kenya where ecological conditions permit, zebu cattle are sometimes replaced by boran cattle. In the semi-arid areas, particularly among the Somali, Boran, and Rendille, camel herding represents an adaptation to the extreme ecological conditions. The animals make quite different demands on pasture: whereas the camels eat the leaves from trees and bushes, cattle live from grass as well as leaves, preferring the former. Camels, moreover, are able to cover long distances, so that the herding families are able to make use of widely separated pastures. Due to natural selection, the animals, in contrast to European cattle, are relatively resistant to diseases and bad feeding conditions. But cattle mortality is very high, the quality of the cattle inferior, and the herds over-aged. Among the herding families the total numbers of cattle are very unevenly distributed. Rational management of the pastoral areas could support a larger population on fewer animals. But cattle not only play a role in the nutrition of the population, they also serve as a form of life assurance, comparable perhaps to the role of the citizen's savings account in the monetary economy. The larger the number of animals, the greater the chance that at least a proportion of them will survive the next drought catastrophe. Another role of the cattle, which should not be underrated, is played in many ceremonies, such as gift, bride-price, and exchange object — features frequently described in the ethnological literature. In this the quality of the animals is of secondary importance, the esteem in which the owner is held being based chiefly on their numbers.

The types of pastoral farming in Kenya are extraordinarily differentiated: from fully developed traditional cattle nomadism via many intermediate forms to modern rotational grazing on cooperative farms, all transitional forms can be found in Kenya. In the semi-arid areas, precipitation is unpredictable and irregular in different regions. After the rains, annual grasses and bushes grow swiftly, but they also dry-up equally speedily when the moisture is expended. In these regions pastoral nomadism is the only form of life and economy adapted to make use of the natural potential. Again and again the nomads must move on the fresh pastures and watering places, with the result that their huts and cattle pens cannot be used continually throughout the year. In some areas old people and children are not able to sustain the long marches and remain in the settlements; often a family's cattle are temporarily split up, with the small animals and a number of milking cows staying in the vicinity of the settlements to supply the remaining members of the family. The able-bodied members migrate with camels or cattle to the remoter areas. The migration routes and customs present the most varied forms. During the rainy season, the herdsmen disperse over wide areas, but with increasing aridity the animals are driven to permanent waterholes. Spencer (1965) has given an impressive account of the Samburu, which can be applied to other herding tribes as well. The result of this behavior is that large areas remain unused, particularly those without waterholes, while the surroundings of waterholes, settlements, and natural salt licks are overgrazed and exposed to soil erosion. In more densely settled areas having more favourable ecological conditions, the herdsmen's migration routes become shorter, and some of them have already adopted a sedentary way of life.

In traditional subsistence pastoral economies there is no personal land ownership, and as a result no-one feels responsible for the care or improvement of pasture. Individualisation of land ownership or the setting up of forms of communal land ownership is the first step towards improvement and modernisation of the pastoral economy. In the Baringo District where the destruction of pastures had reached the worst forms in Kenya already before World War II, the administration introduced modern forms of rotational grazing.

The marketing of cattle is a difficult task, for the animals must be driven over long distances through semi-arid areas. To this end, tracks and enclosures for the overnight accommodation of the animals must be set up. Cattle epidemics such as foot and mouth disease have repeatedly obliged the veterinary authorities to enforce quarantine in certain areas so that the transportation of animals is stopped to avoid infection along the route. On the long march to the abattoir, the animals lose weight, and there is a great risk that a number will die on the way. At present only about 9 per cent of the nomads' animal stocks — as compared with 19 per cent of those from the large farms with a pastoral economy — are sold or consumed annually. The beef cattle supplied by the large grazing enterprises had an average dead weight of 210 kg in the period 1959 – 64, whereas those from the areas of subsistence pastoral economy reached a dead weight of 126 kg only (Republic of Kenya: Development Plan 1966 – 1970, p. 134). The Kenya Government, has recognised that in the development of the subsistence pastoral economy great reserves for the development of the country to be found. The development plans have earmarked large investment sums for the development of the pastoral subsistence economy.

In contrast to areas which are densely settled or recently opened up, little is known concerning the occurrence of disease in these regions. Hospitals exist only at a few administrative centres; dispensaries and health centres are located in small market settlements and do not maintain statistics which could be evaluated. The population is not only very sparse, it is also extremely mobile, so that statements of an epidemiological or geomedical sort mean very little, since they are based on observations carried out in hospitals. In relation to area, the density of the medical services network is extremely low. Systematic observation of epidemic encroachments from contiguous and similar regions in neighbouring states has only begun very recently, as in the case of the introduction of cholera and smallpox in the early seventies.

It must be assumed that the pattern of morbidity and mortality proceeds more or less uninterruptedly or even unobserved by modern health services. Although they alert measures of public aid in drought periods followed by loss of cattle, hunger, and water shortage, with increased mortality in turn (especially among infants, children, and old people) they fail to produce other developments in what is a marginal region as far as human existence is concerned. So too, population figures — to the degree that they can be regarded as moderately reliable under these conditions — do not increase significantly but rather decrease in places. This may be either the result of a high mortality risk or emigration. A disease of particular interest among the Turkana of Lake Rudolf which should be mentioned is echinococcosis (hydatid disease, see Chapter C V, 3).

In these regions trachoma and other infectious eye diseases followed by blindness are very common; these include meningococcal meningitis, although not in the epidemic form so dreaded in the Sahelian Zone of West Africa. Yaws, once widely spread, now only occurs in the north. The Masai are the ones still most accessible from an acculturation away from pastoral life, the first consequence of which appears to be the striking increase in syphilis and gonorrhoea. Diseases which are associated with animal husbandry and the consumption of cattle that have died from natural causes, like anthrax and brucellosis, occur frequently. Among the intestinal helminths, *Taenia saginata* (Beef tape worm) dominates over all others. Malaria occurs naturally in these zones, albeit only sporadically at watering places, which may in turn — since the population lacks adequate immunity or premunity — produce local epidemics. In its visceral form leishmaniasis (Kala Azar) occurs sporadically in the northern and north-western arid areas at least. Tuberculosis definitely poses a considerable problem, whereas leprosy appears to play only a minor role.

2. Class Societies

a) Afro-European class societies engaged in pastoral economy in arid areas: Although large pastoral farms may occur in areas of rain-fed agriculture where mixed farming is possible or is actually carried out, by far the greatest number and acreage of ranches is found in those areas where agriculture is not feasible for climatic reasons. One of these is the Laikipia Plateau, another the area of pastoral farming north and south of Lake Naivasha in the East African Rift Valley. The amount of precipitation in these areas reaches an annual mean of less than 700 to 800 mm over a period in the main less than three humid months. The basis of foodstuffs for the cattle is the natural vegetation, supplemented in only a few, intensively managed, farms by the cultivation of green fodder based on irrigation. According to the ecological map drawn up by Pratt, Greenway and Gwynne (1966), by far the greatest part of the ranches is located in ecological zones 4 and 5. In ecological zone 4 vegetation consists of "dry forms of woodland and savanna (often on *Acacia-Themeda* association), or equivalent bushland". In ecological zone 4 less than 4 ha of grazing are required for the rearing of one unit of cattle. But the carrying capacity of the pasture is greatly reduced by the overgrowth of bush. Pasture also reacts sensitively to over-grazing. Ecological zone 5 is characterized by Pratt et al. (1966, p. 371) as follows: "The woody vegetation being dominated by *Commiphora, Acacia* and allied genera, often of shrubby habit. Perennial grasses such as *Cenchrus cillaris* and *Chloris rox burghiana* can dominate, but succumb readily if the ranch is managed harshly". Here more than 4 ha are needed to support one unit of cattle. A rational utilisation of the grazing potential requires rotational grazing over large areas. Under these ecological conditions and management form on the large pastoral farms the population density must be low indeed. In the Laikipia Plateau it is less than 10 persons per square kilometre, and in the contiguous areas to north and south it even falls below 4 persons per square kilometre. So too, in the East African Rift Valley the population density is very low. Around Lake Naivasha it is below 20 and between Lake Naivasha and Lake Nakuru below 10 persons to the square kilometre.

The farm units are extraordinarily large; they are still almost exclusively owned by Europeans, although their operation depends on African labour. The Africans live in small farm settlements, which, due to the size of the farm units are far away from one another. The living standard of the stockmen on the farms is low, their sanitary conditions are as a rule poor, and, as almost everywhere among the Kenyan Africans, their staple food is maize. The health services are poor, the vast distances within the areas making access to the scanty medical facilities very difficult (see Chapter B, 5).

The large pastoral farms have by no means fully exploited the natural potential. Only about half the grazing is divided up into enclosed pastures and the remaining half is not even made use of everywhere. There are great reserves to be found here, probably on account of the lack of capital or the absence of markets sufficient for the proper development of these farms. On the other hand, some areas with large pastoral farms have seen the pastures overgrown by bush encroachment. This phenomenon of bush invasion Troup (1953, p. 15) explained as resulting from the prohibition of the annual burning, the annihilation of game, the small number of goats, and in some cases over-grazing.

Rotational grazing, bush clearance, seeded grass, fodder cropping, and possibly, if irrigation proved favourable, even hay and silage making—all offer possibilities for improving the animals' feeding basis and the realisation of any significant improvement in the carrying capacity of Kenya's pastoral areas. The supply of animal protein to the growing population urgently requires fast and energetic development of pastoral farming in Kenya.

In a climatically favourable area above the altitude at which malaria is endemic, health conditions are good. The disease pattern is consequently determined more by the ubiquitous infectious diseases like tuberculosis, poliomyelitis, and hepatitis than by specifically tropical ones. Among the farming community, however, a life style increasingly orientated towards the monetary economy increases the danger of malnutrition among children in circumstances under which neither sufficient production for domestic consumption nor much inclination to buy any additional and necessary foodstuffs are observable.

b) Afro-European class societies on large farms in areas of rainfed agriculture: The term Afro-European class society is only valid within certain historic limits, because since 1960, when permission was extended to non-Europeans to acquire farms in the former White Highlands, more and more Africans have bought former European farms. It may be assumed that the Kenyanisation of African farming society will be complete in the near future. This affects the social structure to a certain degree, since not all farms are being continued as large enterprises. To what degree large farms will in the future be divided up and continue to be run as medium-sized or small holdings cannot be stated with any certainty at present.

The core areas of the large-scale mixed farming areas are situated in the Rift Valley in the Nakuru District at about 1,800 m above sea-level and to the west of the Rift Valley. A large and continuous area extends across a pre-Cambrian erosion surface on the Uasin-Gishu and Trans-Nzoia Plateau, in part covered by Tertiary deposits, which was raised to altitudes of 2,100 – 2,700 m at the time of the formation of the fault-block depression. The large-scale mixed farming areas reach their highest altitudes in the region of the Mau Escarpment, where the farms extend above the cultivation limit at an altitude of almost 3,000 m.

The ecological conditions are accordingly varied, but they have one thing in common: namely, they suit Europeans well. The climate manifests the typical characteristics of high altitudes in East Africa: a high degree of solar radiation, moderate temperatures, the diurnal fluc-

tuations of which are greater than their annual fluctuation, low pressure, cooling winds, and precipitation conditions favourable to agriculture. Only at extreme altitudes, as on Mt. Elgon and especially in the areas of the Mau Escarpment, does the climate remain distinctly cool and abundant in rain and mist.

The climatic conditions permit farming in the form of both grain production and animal husbandry, but they also favour the increase of numerous pests, which attack plants and animals and hinder agricultural production. Attacks by locusts and army worm, over large areas require aerial counter-measures. To counter the attacks of wheat rust, new rust-resistant strains must constantly be bred anew. Once a week all the cattle pass through specially installed equipment to be sprayed with insecticides against disease-carrying ticks (Hecklau 1968, p. 169).

As an adaptation to the ecological conditions, areas with mixed farming at an altitude of about 1,800 m above sea-level grow maize, and those at higher levels up to almost 3,000 m grow wheat as their most important grain crop. At altitudes between about 2,000 and 2,500 m pyrethrum constitutes the special crop. In some regions at medium altitudes the farmers grow coffee in relatively small plots. In most areas of mixed farming, the proportion of arable land is about one-quarter to one-third that of all utilisable land on the farms, the remainder laying fallow and serving as natural grazing within the system of extensive pastoralism. Although it had been accepted as far back as the years before World War II that nothing but well-regulated cereal and grassland rotation, together with animal husbandry, would guarantee an optimal utilisation and maintenance of the yield capacity of the soils, farmers were only some years later in a position to put this knowledge into effect. In the fifties, particularly, considerable efforts were made to develop the farms and to switch from one sided cereal growing to mixed farming. By contrast with earlier times, dairy farming now plays a significant role in very many farms. The most intensive form of dairy farming is connected with the cultivation of fodder crops for the additional feeding of fodder concentrates to European breeds of cattle. On the less intensive farms, animal husbandry is confined to the keeping of indigenous breeds, more or less improved by cross-breeding with imported stock. The centre of sheep raising in the areas of mixed farming is to be found in the highlands of the Molo-Mau-Narok District, where sheep provide one of the main sources of income on the farms. Elsewhere they play only a modest role.

The so-called White Highlands have never really been the land of the white man, not even in the mixed farming districts. This can be demonstrated in the case of Trans-Nzoia and Uasin Gishu Districts. Taking the number of Europeans of both districts round about 1960, that is at the zenith of the White Highland's development by the Europeans, to be about 3,000 it would mean a population density of six Europeans per square kilometre. But as long as the White Highlands existed, unwanted but irresistably Africans from contiguous and densely populated native reserves penetrated the White Highlands, settling where they were not definitely stopped. In 1962, the African population had grown to almost 200,000 persons, the Africans outnumbering the Europeans by 80 to 1. Today the population density in mixed farming districts runs to about 100 inhabitants per square kilometre, although it remains less than 20 per square kilometre in some areas. It is therefore understandable that in the face of low population density and the by no means fully utilised agricultural potential, the wish has arisen among many Africans to divide the large farms and to settle the land more densely. "The One Million Acre Settlement Scheme" (cf. Chapter D 4) may thus possibly represent only the beginning of a development, the end of which will see the disappearance of the large-scale mixed farms.

c) Afro-European class societies on coffee plantations: Coffee is one of Kenya's most important export crops, although most coffee is produced on large-scale mixed farms and small-scale African farms. It cannot be determined exactly how many farms are exclusively coffee producers, and might thus be described as plantations.

"Thus the six biggest companies have rather less than 12,000 acres (=4,856 ha) of coffee (under 6 per cent of the whole), while the remaining 68,000 acres (=27,518 ha) of estate coffee (about one third of the whole coffee acreage) are in the hands of an unknown number of private companies and individuals, still probably largely non-Kenyans. The capital involved in the coffee industry is thus mainly local, and little large-scale investment has been attracted from overseas" (Who controls industry in Kenya 1968, p. 16 et seq.).

Coffee plantations are to be found mainly on the eastern edge of the Aberdares, whereas in other areas of the former White Highlands the cultivation of coffee usually remains a secondary branch of the mixed farming activity.

In the Aberdares the arabica coffee plant encounters ideal conditions for growth. Here the volcanic and slightly acid soils are deep and rich in nutrients. They are distinguished by their good drainage. The average annual precipitation amounts to between 800 and 1,100 mm. Dissected by deep valleys, the area presents numerous perennial streams, and water for the processing of the coffee berries is available throughout the year. Labour is relatively easily recruited from the neighbouring and densely settled Kikuyu District. The closeness of Nairobi, the capital, facilitates the organisation of the production and sale of coffee. For these reasons the only continuous area of coffee plantations in Kenya might well develop here. But because of the lack of suitable land, it remains restricted to a narrow belt, which extends between the densely settled Aberdares and the arid pastures and sisal growing areas of Thika. Originally the area belonged to the settlement area — or at least the sphere of influence — of the Kikuyu, and in no other part of Kenya did the interests of the Africans and Europeans clash to such a degree and with such severity (Hecklau 1970, p. 193).

d) Afro-European class societies on tea plantations: In Kenya the ecological conditions suitable for tea cultivation occur only at altitudes of between 1,500 and 2,400 m where the annual precipitation amounts to at least 1,300 mm and falls as regularly as possible throughout the year. Soils must be slightly acid and moist, but must be free from stagnant water. Apart from a number of small tea plantations in various parts of Kenya, tea cultivation is concentrated in the Kericho District, where in 1971 about 12,000 ha were under tea cultivation, in the Nandi District, about 5,000 ha of tea, in the North Sotik District, 1,335 ha and, finally, in the Limuru District, 1,659 ha. Without doubt the largest part of the tea acreage is under the ownership of a few large commercial tea companies. In order to give the African farmers a share in tea cultivation in Kenya as well, the Kenya Tea Development Authority was founded in 1964.

However, the production of tea depends on technical conditions requiring great capital outlay. Tea bushes only produce their first crop 4 to 6 years after planting, and during this period labour-intensive care must be devoted to them. A certain minimum acreage — in Kenya around 150 ha — is required for full technical utilisation of a tea factory. Such a factory must be located on the plantation or in its vicinity, since the plucked tea leaves spoil within a few hours if they are not processed. In those tea growing areas which enjoy a lot of rain, the plantations must be linked by a road network suitable for all-weather traffic. It is calculated that on the average one labourer is needed for every 0.4 ha of tea in a plantation having its own tea factory. Plantation workers normally live with their families in housing owned by the plantation company. A 150 hectare tea plantation thus requires housing for about 1000 persons, added to which will be a school and a dispensary, as well as a building for the administration. Last and not least comes the most costly investment, the tea factory itself: such a factory cost around 1.0 to 1.5 million DM in the late sixties. These high investments can only be afforded by companies with sound capital assets. The form of management is usually a commercially run, monocultural plantation. Individual farmers can take part in tea production only if they are able to sell their leaves to the centrally located tea plantations of factories.

The economy of tea plantations, particularly in its vast expansion over much of the Kericho District, has completely altered the physiognomy of the landscape. Instead of the montane rain forest which once characterised the natural landscape here, there is now a well cared for cultural landscape, with neatly clipped and bright green tea plantations. Standing over them at regular intervals are the shade trees which run throughout the plantations. The clean housing of the plantation workers' families, surrounded by lawns, thus takes the form either of whitewashed houses with thatched, conical roofs in the manner of African round huts, or of small square houses, regularly aligned and well spaced. At the centre of the plantation and convenient for transport are the tea factory and the administrative buildings.

Although the so-called "labour lines", the settlements of the plantation workers, give the impression of a smart appearance, at least externally and the labourers are in possession of a regular income, their standard of domestic hygiene and nutrition is by no means what it could be under the prevailing circumstances. Only some minimal requirements, such as latrines and water-supply for the family instead of collective usage, health education and dietetics linked with the handling of cash, and emphasis on pre-natal and child care in a situation of prevalent female labour, can guarantee adequate standards of life and health. This is shown by an example from Ceylon, where E. Kröger (1970) was able to demonstrate that infant mortality was higher and the state of health of women on the tea plantations worse than in the surrounding population, because these minimal requirements were not available.

e) Afro-European class societies on sisal plantations: Before the invention of artificial fibres, sisal was an ideal plantation crop, for sisal can flourish in those areas which are little if at all suited to mixed farming. Neither prolonged dry periods nor wet years, neither hot spells nor pest attack seem able seriously to harm this persevering plant. It flourishes even on poorer, stoney soils, although in terms of quantity and quality better yields are grown from good soils in a warm and humid climate. A further locational precondition is the existence of water, great quantities of which are needed for decortification of the sisal. Although the demand for sisal has always been subject to great fluctuations, the modern development of superior and cheaper artificial fibres constitutes a serious threat to sisal growing.

"The production of sisal is very largely in the hands of about fifty five estates, totalling 250,000 acres, which produced 55,100 tons in 1966, dropping to 50,600 tons in 1967 . . . the sisal estates fall into two groups. About thirty of them are large, including a few giants, and up to 1966 they produced over 1,000 tons a year each. This group accounted for 50,000 tons a year. The remaining twenty five estates are smaller, and the average annual production is about 500 tons each" (Who controls industry in Kenya, 1968, p. 25).

The lay-out of sisal plantations has a great spatial effect in Schultze's opinion (1966, p. 2). Like all large-scale plantations laid out in a natural landscape thus far barely altered by man, they bring about a profound landscape change. The labour intensive cultivation of sisal requires the settlement of great numbers of people in previously scarcely inhabited regions. The labourers come from distant peasant farming areas characterised by great population density but few employment opportunities. A modern and well-organised sisal plantation requires a workforce of more than 1,000 persons. The inaugurators of sisal plantations have not only to build accommodations for their workers, but also to provide roads and tracks, bridges, loading facilities, storage sheds, factories, workshops, hospitals, churches and mosques, as well as markets and shops for the supply of the population (Hitchcock 1959, p. 7).

In the fifty or so sisal plantations in Kenya, an estimated total of 25,000 workers are employed. According to information provided by the Kenya Sisal Board in Nairobi, these are supplemented by some thousands of workers in the sisal factories. If the non-working members of their families are included, more than 100,000 people may depend for their existence on the sisal industry. In 1971 the cultivation of sisal was chiefly carried on in the Nakuru District (17,848 ha), the Taita District (21,803 ha), the Machakos District (16,622 ha) and in the Muranga, Kilifi, and Kiambu District, each of which totalled between 5,000 and 10,000 ha of sisal cultivation. On account of the difficulties in the market for sisal which results from the competition of synthetic fibres, sisal production is declining.

Apart from health and health service problems characteristic of plantations (but quite likely severe in large-scale sisal plantations, although not confirmed by detailed information), there is a problem of massive schistosomiasis infection on the sisal plantations of the coastal area, and especially in the Taveta-Taita District. This problem might be solved better than in distinct irrigation projects, insofar as water is not a production input in the same way here, nor is barely avoidable contact an occupational necessity for the labourers and population present here to the same degree.

f) Afro-Asian class societies on sugar plantations: Sugar cane on plantations is grown only in the Kavirondo Gulf, east of Kisumu, and in the coastal area near Ramisi. Here the sugar cane encounters ideal growing conditions: "Compared with some other areas of sugar cane cultivation, the East African ones enjoy the advantage that as a result of their relatively constant climatic conditions and the feasibility of artificial irrigation, almost all the year round harvesting, as well as planting of cuttings, is possible. This thus permits a continuous manufacturing process in the cane sugar refineries and the full utilisation of productive capacity" (Engelhard and Lienau 1970, p. 55). Whereas the production of tea, coffee, and sisal depends on export possibilities, Kenyan sugar production does not even meet half the domestic demand. Production has been increased considerably over the period 1968–71, rising from about 19,000 to 28,000 ha. Rather more than half the area cultivated is owned by large sugar factory companies, about one-third of the sugar cane is grown by large farms in the vicinity of the sugar factory companies, and the remainder is produced by smallholders. The large farms and the smallholders supply the sugar cane to these companies. The processing of cane sugar is controlled by a few large firms, most of which are under the ownership of the Miwani, Madhvani, and Metha concerns, namely enterprises owned by Asian families. By means of the Chemilil Sugar Scheme and the erection of a sugar refinery, the state is attempting to improve the living conditions of the peasant population by giving them the opportunity to earn cash through the growing of sugar cane and its sale direct to these two factories.

Whereas mass administration of drugs to the 3,000 settlers has been successfully carried out on the Ramisi Sugar Estates by the Department of Vector-Borne Diseases since 1966 in order to counter the high level of malaria morbidity, which interfered with the project's development, the problem was not considered to be significant in the Chemilil Sugar Scheme. This was probably because malaria was in any case endemically stable there, with the result that there were no production losses.

Conclusions

Vast, gently undulating plains combined with the high ranging Kenya Highlands crowned by snow-capped Mount Kenya and intersected by the impressive Rift Valley system — these are the main physiographical features of Kenya. The strikingly diverse landforms in very different altitudes above sea level due to Kenya's location below the equator on the eastern part of the African continent determines the climate of the country; the relief is particularly decisive for the quantity and distribution of rainfall, as well as for the temperature levels. These factors above all others in their turn determine the ecological conditions of a predominantly agrarian society. The ecological endowment is the basis for the distribution of the population and the regional development potential.

Inevitably linked with the climatic, ecological and economic conditions are the occurrence and distribution of diseases and disease vectors as well as the distribution of and the accessibility to health services, of paramount importance for the development potential of the country.

Rapid and uneven growth of the population and the economy have differentiated the more or less egalitarian pastoral- and peasant society of pre-colonial times. The development of a more diversified economy has led to the opening up to the entire country by means of communication linkages to an unforeseen extent. In turn, the population has attained a high level of mobility. Changes in all spheres of life — in education, habitation, nutrition, domestic hygiene, daily economy, modernisation of rural settlements together with the change in form and building material, as well as urbanisation have modified the traditional disease pattern and thus in turn greatly influenced the mortality pattern and the population growth rate as much as, if not more than, the increasing availability of curative and preventive health services.

The growth of the population and the increasing scarcity of land available for agricultural use is causing more people to move from rural to urban areas in want for job-opportunities and employment in the non-agricultural sectors. This results in a deterioration of the standard of living and housing conditions with the typical concomitant social and medical problems.

About 90 per cent of the Kenyan population lives directly or indirectly from agriculture, the production conditions of which depend upon ecological factors. In particular, pre-industrial societies have to adapt themselves more closely to the ecological conditions of their habitat than is the case of an industrial society; thanks to technological means and know-how, the latter are better equipped to overcome unfavourable natural conditions more efficiently. Nevertheless, in agricultural societies the natural ecological conditions provide only the framework for man's activities. To which degree and in which manner this framework is expanded depends on socio-economic conditions and their historic roots.

A decisive epoch in the economic and social history of Kenya can be recognised in the period of large farm, ranching, and plantation development, and associated with it the commercialisation of erstwhile subsistence-orientated African agricultural and pastoral economies.

Over recent decades highly differentiated agrarian societies have developed within what is ecologically speaking, the extraordinarily diversified territory of Kenya; these have used the ecological potential of their habitats in a variety of ways. The conditions of life and the living standard of the Kenyan are determined by the physico-geographical endowment of his habitat as well as by membership of a particular social group.

Keeping in mind these facts, the attempt was made to divide the territory of Kenya into regions in which the ecological conditions and the socio-economic circumstances of the population show features different from those of neighbouring regions.

This Geomedical Monograph of Kenya chooses the ecological approach to the question of causation and prevention of diseases, looking at the entire environment, the biological and natural sphere as well as at the socio-economic complex. Geomedical analysis of the disease distribution within a population and geomedical synopsis of an area are two semi-quantitative means of looking at the natural and man-made history of disease.

Physico-geographical conditions, demographic, economic and socio-cultural structures, on the one hand, and the prevalence and incidence of disease, on the other, are examined for their spatial correlation in order to find indications for possible causal relationship in medical-ecological terms. Special attention is given to the structural and functional analysis of the health services and to the presentation of the occurrence of the most important diseases, as they appear in their temporal and regional dimension. On the basis of existing data and special investigations, it was possible to distinguish fifteen more or less clearly differentiated economic and socio-geographical regions, which formed the basis of the geographical and geomedical synopsis.

This was done by presenting the conditions of life as the outcome of inter-relationships between ecological, economic, socio-geographical factors and geomedical and epidemiological factors, taking inherent methodological problems into due account.

This interdisciplinary approach may help towards a better understanding of the very complex process of development and may serve as an example for geomedical analyses of other developing countries.

Annex: Tables

Table I. Population according to tribes

Major Group	Tribe	Inhabitants 1969	
Central Bantu	Kikuyu	2,201,632	
	Embu	117,969	
	Meru	554,256	
	Mbere	49,247	
	Kamba	1,197,712	
	Tharaka	51,883	4,172,699
Western Bantu	Luhya	1,453,302	
	Kisii	701,679	
	Kuria	59,875	2,214,856
Coastal Bantu	Mijikenda	520,520	
	Pokomo/Riverine	35,181	
	Taveta	6,324	
	Taita	108,494	
	Swahili/Shirazi	9,971	
	Bajun	24,387	
	Boni/Sanye	3,972	708,849
Nilotic	Luo	1,521,595	1,521,595
Nilo-Hamitic: (Kalenjin-speaking	Nandi	261,969	
	Kipsigis	471,459	
	Elgeyo	110,908	
	Marakwet	79,713	
	Pokot	93,437	
	Sabaot	42,468	
	Tugen/Cherangani	130,249	1,190,203
Other Nilo-Hamitic	Masai	154,906	
	Samburu	54,796	
	Turkana	203,177	
	Iteso	85,800	
	Nderobo	21,034	
	Njemps	6,526	526,239
Western hamitic (Rendille and Galla-speaking)	Rendille	18,729	
	Boran	34,086	
	Gabbra	16,108	
	Sakuye	4,369	
	Orma	16,306	89,598
Eastern hamitic (Somali-speaking)	Gosha	2,926	
	Hawiyah	4,103	
	Ogaden	90,118	
	Ajuran	15,544	
	Gurreh	49,241	
	Degodia	62,425	
	other Somali	25,374	249,731
			10,673,770

Source: Republic of Kenya: Kenya population census, 1969, Vol. I, p. 69

Table II. Small Farms and Settlement Schemes *
Estimated Crop Areas by Rains Cycle

'000 Hectares

	1969 Short Rains			1970 Long Rains			Crop Year 1969/70		
	Single Crop	Mixed Crop	Total	Single Crop	Mixed Crop	Total	Single Crop	Mixed Crop	Total
Cereals									
Improved Maize	13.1	10.3	23.4	86.2	37.8	124.0	99.3	48.1	147.4
Unimproved Maize	68.0	290.4	358.4	107.9	382.0	489.9	175.9	672.4	848.3
Bulrush Millet	1.8	28.6	30.4	1.7	12.6	14.3	3.5	41.2	44.7
Finger Millet	2.6	5.7	8.3	10.1	17.7	27.8	12.7	23.4	36.1
Other Miller	0.2	6.7	6.9	1.7	4.5	6.2	1.9	11.2	13.1
Sorghum	6.3	30.3	36.6	18.7	85.9	104.6	25.0	116.2	141.2
Wheat	3.5	—	3.5	2.0	—	2.0	5.5	—	5.5
Other Cereals	0.3	0.3	0.6	0.6	2.1	2.7	0.9	2.4	3.3
Pulses									
Beans	7.2	147.7	154.9	2.3	165.4	167.7	9.5	313.1	322.6
Pigeon peas	—	53.6	53.6	0.5	67.9	68.4	0.5	60.7	61.2
Cow peas	0.1	30.7	30.8	0.1	35.7	35.8	0.2	66.4	66.6
Field peas	0.8	7.0	7.8	—	5.0	5.0	0.8	12.0	12.8
Yellow, Green and Black Grams	—	3.8	3.8	0.4	10.2	10.6	0.4	14.0	14.4
Other Pulses	—	0.7	0.7	—	1.3	1.3	—	2.0	2.0
Temporary Industrial Crops									
Cotton	22.5	17.0	39.5	15.3	10.9	26.2	37.7	27.9	65.6
Sugar cane	10.2	15.9	26.1	12.0	18.4	30.4	11.1	17.2	28.3
Pyrethrum	10.5	1.0	11.5	11.0	8.3	19.3	10.7	4.6	15.3
Groundnuts	0.4	4.9	5.3	2.3	14.4	16.7	1.3	9.6	10.5
Oil Seeds	2.0	13.9	15.9	—	8.1	8.1	1.0	11.5	12.5
Other Temporary Industrial Crops	1.5	3.0	4.5	1.2	6.4	7.6	1.3	4.7	6.0
Other Temporary Crops									
Cassava	18.6	50.4	69.0	12.1	72.4	84.5	15.4	61.4	76.8
English Potatoes	5.1	11.9	17.0	5.0	15.6	20.6	10.2	27.5	37.7
Sweet Potatoes	3.6	13.2	16.8	4.0	24.6	28.6	3.8	18.9	22.7
Yams	0.2	7.7	7.9	0.5	7.8	8.3	0.4	7.8	8.2
Cabbages	2.7	4.7	7.4	0.3	4.7	5.0	3.0	9.4	12.4
Other Vegetables	—	0.6	0.6	0.3	0.3	0.6	0.3	0.9	1.2
Other Temporary Crops	2.2	12.0	14.2	6.6	7.6	14.2	8.8	19.6	28.4
Permanent Crops									
Coffee	—	—	—	51.1	11.4	62.5	51.1	11.4	62.5
Tea	—	—	—	19.3	0.3	19.6	19.3	0.3	19.6
Coconuts	0.6	34.9	35.5	0.5	43.9	44.4	0.6	39.4	40.0
Cashew nuts	1.3	29.7	31.0	1.1	36.6	37.7	1.2	33.2	34.4
Bananas	9.5	59.8	69.3	10.7	70.9	81.6	10.0	65.3	75.3
Other Fruit	1.2	20.4	21.6	1.5	23.2	24.7	1.3	21.8	23.1
Other Permanent Crops	2.2	1.4	3.6	3.8	1.5	5.3	3.0	1.5	4.5
Total cultivation	198.2	411.8	610.0	390.9	546.2	937.1	589.1	958.0	1,547.1

Source: Statistical Abstract 1975 Table 86

* The following districts are excluded from this table:
Nakuru, Laikipia, Uasin Gishu, Trans Nzoia, Baringo, West Pokot, Lamu and Tana River. The first four are predominantly large farm areas and crop areas for these are included in the data for large farms.

Table III. LARGE FARMS Land Utilization by District, 1974   Hectares

| | Nyanza Province | | | Rift Valley Province | | | | | | | |
	Kisumu	Kisii	Total	Baringo	Kericho	Trans Nzoia	Laikipia	Nakuru	Nandi	Uasin Gishu	Total
Cereals											
Wheat	—	—	—	—	2,316	4,960	2,990	29,112	—	34,837	74,215
Maize	7	300	307	—	2,754	23,993	2,811	7,765	447	23,128	60,898
Barley	—	—	—	—	611	187	1,032	6,719	—	711	9,260
Oats	—	—	—	—	70	460	238	2,605	6	1,219	4,598
Others	—	4	4	—	25	181	63	644	—	176	1,089
Total	7	304	311	—	5,776	29,781	7,134	46,845	453	60,071	150,060
Temporary Industrial Crops											
Sugarcane	13,377	—	13,377	—	360	2	—	77	6,872	71	7,382
Pyrethrum	—	11	11	17	312	4	101	3,345	10	194	3,983
Sunflower	—	13	13	—	4	4,567	61	540	41	469	5,682
Others	—	—	—	—	—	—	—	342	1	152	495
Total	13,377	24	13,401	17	676	4,573	162	4,304	6,924	886	17,542
Root Crops and Vegetables	—	4	4	—	19	233	258	752	36	144	1,442
Temporary Fodder Crops	—	1	1	—	39	1,784	684	4,941	56	313	7,817
Other Temporary Crops	70	466	536	92	3,550	7,528	6,072	20,413	1,663	8,679	47,998
Temporary Meadows and Fallow land	893	107	1,000	57	4,126	32,069	15,078	39,500	1,604	21,051	113,485
Permanent Industrial Crops											
Sisal	—	—	—	—	—	—	—	21,755	—	2,234	23,989
Tea	2	925	927	—	14,180	182	210	235	5,810	1,519	22,136
Coffee	—	5	5	—	686	1,316	95	2,440	166	58	4,761
Wattle	—	—	—	—	4	65	4	123	—	12,776	12,972
Coconut	—	—	—	—	—	—	—	—	—	—	—
Others	—	—	—	—	828	9	—	28	9	226	1,100
Total	2	930	932	—	15,698	1,572	309	24,581	5,985	16,813	64,958
Fruit											
Cashew nuts	—	—	—	—	—	—	—	—	—	—	—
Pineapples	—	—	—	—	—	10	—	—	—	2	12
Others	—	—	—	—	9	175	5	174	—	20	383
Total	—	—	—	—	9	185	5	174	—	22	395
Uncultivated Meadows and Pastures	4,178	4,147	8,325	6,057	55,456	107.024	705,560	314,520	29,565	180,390	1,398,572
All Land	18,527	5,983	24,510	6,223	85,349	184,749	735,263	456,030	46,286	288,369	1,802,2692

	Western Province			Central Province				
	Bungoma	Kakamega	Total	Kiambu	Nyandarua	Nyeri	Murang'a	Total
Cereals								
Wheat	—	—	—	117	3,726	2,671	—	6,514
Maize	55	13	68	253	460	329	922	1,964
Barley	—	—	—	—	1,343	106	—	1,449
Oats	—	—	—	5	305	30	1	341
Others	—	—	—	4	3	—	—	7
Total	55	13	68	379	5,837	3,136	923	10,275
Temporary Industrial Crops								
Sugarcane	—	3,641	3,641	1	—	—	—	1
Pyrethrum	—	—	—	1	241	1	—	243
Sunflower	44	—	44	13	—	1	4	18
Others	—	—	—	—	—	—	—	—
Total	44	3,641	3,685	15	241	2	4	262
Root Crops and Vegetables	—	—	—	300	133	92	159	684
Temporary Foldder Crops	—	—	—	378	1,029	195	60	1,662
Other Temporary Crops	38	13	51	9,992	6,748	790	5,126	22,656
Temporary Meadows and Fallow Land	13	—	13	5,053	7,713	3,159	505	16,430
Permanent Industrial Crops								
Sisal	—	—	—	2,302	—	—	7,230	9,532
Tea	—	—	—	3,282	—	—	—	3,282
Coffee	13	—	13	14,059	—	881	5,431	20,371
Wattle	—	—	—	475	5	102	14	596
Coconut	—	—	—	—	—	—	—	—
Others	—	—	—	87	4	—	6	97
Total	13	—	13	20,205	9	983	12,681	33,878
Fruit								
Cahew nuts	—	—	—	—	—	—	2	2
Pineapples	—	—	—	2,964	—	—	225	3,189
Others	—	—	—	165	2	2	169	338
Total	—	—	—	3,129	2	2	396	3,529
Uncultivated Meadows and Pastures	640	1,231	1,871	43,851	114,545	40,734	34,831	233,961
All Land	803	4,898	5,701	83,302	136,257	49,093	54,685	323,337

Table III. (Contd.) LARGE FARMS Land Utilization by District, 1974   Hectares

	Eastern Province			Coast Province				Nairobi	Total
	Machakos	Meru	Total	Kilifi	Kwale	Taita	Total		
Cereals									
Wheat	—	8,522	8,522	—	—	—	—	—	89,251
Maize	374	38	412	10	—	—	10	62	63,721
Barley	—	1,371	1,371	—	—	—	—	11	12,091
Oats	—	74	74	8	—	—	8	—	5,021
Others	3	—	3	—	—	—	—	1	1,104
Total	377	10,005	10,382	18	—	—	18	74	171,188
Temporary Industrial Crops									
Sugarcane	130	137	267	—	4,674	—	4,674	—	29,342
Pyrethrum	—	14	14	—	—	—	—	—	4,251
Sunflower	—	—	—	—	—	—	—	6	5,763
Others	—	—	—	—	—	—	—	6	501
Total	130	151	281	—	4,674	—	4,674	12	39,857
Root Crops and Vegetables	321	43	364	15	91	—	106	125	2,725
Temporary Fodder Crops	416	1,872	2,288	55	4	—	59	100	11,927
Other Temporary Crops	3,756	147	3,903	34	25	—	59	547	75,750
Temporary Meadows and Fallow Land	13,899	8,604	22,502	1,385	526	20,351	22,262	324	176,016
Permanent Industrial Crops									
Sisal	16,727	—	16,727	8,077	—	22,394	30,471	1,136	81,855
Tea	—	—	—	—	—	—	—	—	26,345
Coffee	1,996	—	1,996	—	—	—	—	1,352	28,498
Wattle	—	4	4	—	—	—	—	—	13,572
Coconut	—	—	—	340	1,375	—	1,715	—	1,715
Others	—	—	—	8	294	—	302	6	1,505
Total	18,723	4	18,727	8,425	1,669	22,394	32,488	2,494	153,490
Fruit									
Cashew nuts	—	—	—	646	173	—	819	—	821
Pineapples	—	—	—	—	11	—	11	—	3,212
Others	28	1	29	55	56	4	115	11	876
Total	28	1	29	701	240	4	945	11	4,909
Uncultivated Meadows and Pastures	202,530	22,786	225,316	7,784	5,728	7,380	20,892	22,018	1,910,955
All Land	240,179	43,613	283,792	18,417	12,957	50,129	81,503	25,705	2,546,817

Source: Statistical Abstract 1975, Table 92

Table IV. Quantity and Value of Fish Landed
Selected Years 1966 – 1972 and Projected for 1978

Area and Main Group		1966	1972	1978
		Tons	Tons	Tons
Fresh Water Fisheries				
Lake Victoria		15,445	15,989	17,500
Lake Rudolf		1,524	4,090	4,750
Lake Naivasha — Commercial		914	117	1,150
Lake Naivasha — Sport		—	144	200
Lake Baringo		610	58	700
Other Lakes		965	208	1,000
Rivers		1,524	1,624	1,600
Fish Ponds		203	—	100
Sub-Total		21,185	22,230	27,000
Marine Fisheries				
Marine Fish		6,507	7,411	7,600
Crustaceans		140	185	200
Other Marine Products		223	126	200
Sub-Total		6,870	7,722	8,000
Total Fish Landed (Tons)		28,055	29,952	35,000
Total Value to Fishermen	K£	1,433,500	1,500,000	1,725,000

Source: Development Plan 1974 – 1978, p. 267

Table V. Visitor Days in Kenya
by Country of Residence of Visitors and Purpose of Visit, 1971,
1972, 1978

('000 *Days*)

Purpose/Residence	1971	1972	1978
Holiday Visitors	Recorded		Target
Residents of:			
Other African Countries	206.5	278	244.2
North America	447.0	270	1,734,0
United Kingdom	546.0	709	1,313.0
West Germany	345.0	469	1,467.0
Other European Countries	473.0	676	1,291.0
Other Countries	247.0	290	588.0
Total Holiday, Non-E.A.	2,264.0	3,092	6,637.0
Business Visitors, non-E.A.	304.0	275	496.5
Transit Visitors, non-E.A.	57.0	52	112.0
Total of Non-E.A. Residents	2,625.0	3,419	7,245.0
Total of E.A. Residents	1,109.0	1,350	1,000.0
Total All Visitors	3,734.0	4,769	8,244.0

Source: Development Plan 1974 – 1978, Part I, p. 381.

Table VI. Details on "Estimates of Recurrent Expenditures" on Health by the Ministry of Health *

K£	1963/64	%	1968/69	%	1973/74	%
Personal Emoluments, Housing, Leave etc.	1,342,886/—	55,96	2,534,330/—	55,50	5,261,572/—	55,37
Travel, Transport, Repair and Purchase of Vehicles	60,465/—	2,56	111,350/—	2,44	239,675/—	3,20
Running Costs of Ministry of Health and Central Services	189,003/—	7,61	228,060/—	4,94	385,137/—	4,06
Training of non-academic medical staff	?	—	116,170/—	2,54	273,152/—	2,87
Grants and Subsidies to non-government medical institutions	57,256/—	2,40	72,285/—	1,50	275,260/—	2,89
Medical Institutions etc.	344,836/—	14,38	371,200/—	8,13	586,810/—	6,18
Medical Stores, Plant, Equipment, Medicaments	400,714/—	16,70	779,300/—	17,07	1,091,460/—	11,48
Mother and Child Health Services	—	—	15,000/—	0,33	178,737/—	1,88
Health Centres and Dispensaries (since 1970) Personnel since 1973/74 included in Personal Emoluments)	—	—	—	—	466,550/—	4,90
National Hospital Insurance Fund	—	—	46,000/—	1,01	62,262/—	0,65
Kenyatta National Hospital (excl. Personnel)	—	—	292,805/—	6,41	678,985/—	7,15
Total Gross Estimates of Recurrent Expenditure Ministry of Health	2,395,160	100,00	4,566,500	100,00	9,499,600	100,00
Per Capita Recurrent Expenditure of Ministry of Health	5,3 Ksh		9,13 Ksh		15,83 Ksh	

* Comparative combined table contrasting the three most important fiscal years of the last decade according to the published "Estimates of Recurrent Expenditure", Government of Kenya, Nairobi 1963, 1968, 1973.

Table VII. "Estimates of Recurrent Expenditures" of the Government of Kenya 1968/69 – 1973/74 *

	Approved Gross Estimates in K£					in relation to Census population estimates		
Fiscal year	Total Expenditure of the Government of Kenya	%Growth against Preceeding Year	Total Expenditure Ministry of Health	% of Total Expenditure of Government of Kenya	% Growth against Preceeding Year	Population Estimate in Million and crude % Growth rate against preceeding year	Central Government per Capita Expenditure on Health Ksh	%Growth against Preceeding Year
1967/68	58,792,064/—	—	3,537,500/—	6.01	—	9.9	7.14	
1968/69	61,321,800/—	4.30	3,847,300/—	6.27	+ 8.76	10.2 = +3.30	7.54	+ 6.00
1969/70	79,485,854/—	29.63	4,981,350/—	6.27	+29.48	10.9 = +6.42	9.14	+21.22
1970/71	108,270,077/—	36.21	6,828,300/—	6.31	+37.08	11.2 = +2.75	12.19	+23.33
1971/72	126,452,866/—	16.83	8,277,850/—	6.55	+21.23	11.7 = +4.46	14.15	+16.08
1972/73	138,889,991/—	9.83	9,490,476/—	6.83	+13.44	12.1 = +3.42	15.68	+10.81
1973/74 **	153,611,344/—	10.60	9,499,600/—	6.18	+ 0.95	12.5 = +3.31	15.19	− 3.23

* Analysis compiled from "Estimates of Recurrent Expenditures" of the Government of Kenya, 1967 to 1974, Government Printer, Nairobi; not included are expenditure by local Governments, Mission hospitals, and Development Expenditures on Health.

** Gross Estimate 1973/74

Table VIII. Hospital and hospital-bed-ratio per province in relation to areas and population to be covered, Kenya 1970

Province	Land area in km²	Population in Million	Percent of total pop.	Pop. density per km²	Total No. of Hosp.	%	Total No. of beds incl. cots	%	1 Hospital per X pop.	1 Hospital per X km² land area	No. of beds per 1,000 pop.
Nairobi extra provincial area	684	0.54	4.7	790	20	11	3,134	21.6	1 : 27,000	1 : 34	5.80
Central	13,173	1.78	15.3	136	30	16.5	2,246	15.5	1 : 59,333	1 : 439	1.26
Coast	83,041	1.00	8.6	13	21	11.5	1,603	11.0	1 : 47,619	1 : 3,954	1.60
Eastern	154,540	2.02	17.4	14	26	14.3	1,997	13.7	1 : 77,692	1 : 5,944	0.99
N.E.	126,902	0.26	2.2	3	3	1.65	106	0.73	1 : 86,666	1 : 42,300	0.41
Nyanza	12,526	2.25	19.4	180	23	12.6	1,474	10.1	1 : 97,826	1 : 545	0.66
Rift Valley	170,163	2.35	20.2	14	47	25.8	2,873	19.8	1 : 50,000	1 : 3,620	1.22
Western	8,223	1.41	12.2	172	12	6.6	1,092	7.5	1 : 117,500	1 : 685	0.77
Total	569,252	11.61	100.0	21	182	100	14,525	100	1 : 63,791	1 : 3,127	1.25

Source: Basic data from Rep. of Kenya, Ministry of Health, Health Services in Kenya, 1971, Nairobi 1973 and population census 1969 (cf. Ann. Rep. 1966, p. 100).

Table IX. Health Centres and Dispensaries by Operating Agency and Province, 1970 and Health Centre — and Dispensary — Population Ratio per Province

Province	Health Centres (HC) Central Govt.	Comp.	Mis-sion	Munic.	Σ	%	HC: Pop.	Dispensaries (D) Central Govt.	Comp.	Mis-sion	Munic.	Σ	%	Disp. pop.	HC : D
Nairobi extra Provincial area	—	—	—	6	6	3.1	1 : 90,000	21	8	4	10	43	7.1	1 : 15,558 +HC 1 : 11,020	1 : 7.17
Central	31	—	—	—	31	15.9	1 : 57,419	54	7	44	—	105	17.4	1 : 16,952	1 : 3.39
Coast	12	—	—	11	23	11.8	1 : 43,478	75	2	4	4	85	14.1	1 : 11,765	1 : 3.70
Eastern	24	—	—	—	24	12.3	1 : 84,166	71	1	39	—	111	18.4	1 : 18,200	1 : 4.63
North-Eastern	3	—	1	—	4	2.0	1 : 65,000	12	—	—	—	12	2.0	1 : 21,667	1 : 3.0
Nyanza	28	—	—	—	28	14.4	1 : 80,357	44	1	23	—	68	11.3	1 : 33,088	1 : 4.43
Rift Valley	50	—	2	1	53	27.2	1 : 44,340	93	40	32	—	165	27.4	1 : 14,242	1 : 3.11
Western	25	—	1	—	26	13.3	1 : 54,230	8	—	6	—	14	2.3	1 : 100,714	1 : 0.54
Total n	173	0	4	18	195	100%	1 : 59,538	378	59	152	14	603	100%	1 : 19,254	1 : 3.1
%	88.7	0	2.0	9.3	100%			62.7	9.8	25.2	2.3	100%			
N				195								603		=798 1 : 14,548	1 : 3.9

Table X. The present strength of medical officers in Government Hospitals

P.G.H. Province, District hospital	Specialist		Medical Officers		Medical Officers (intern)		Total	
	n	%	n	%	n	%	n	%
Kenyatta National Hospital, Nairobi	31	52	57	49.5	90	37.8	178	43.1
P.G.H. Mombasa	7	12	10	8.7	14	5.9	31	7.5
Coast Prov.	—		—					
9 Distr. Hospitals	—		—		11		11	2.6
P.G.H. Machakos	4	6.6	8	6.9	10	4.2	22	5.3
Eastern Prov.								
8 Distr. Hospitals	—		—		15	6.3	15	3.6
P.G.H. Nyeri	4	6.6	10	8.7	10	4.2	24	5.8
Central Prov.								
8 Distr. Hospitals	—		—		17	7.1	17	4.1
P.G.H. Nakuru	5	8.3	10	8.7	12	5.0	27	6.5
Rift Valley Prov.								
18 Distr. and Subdistr. Hospitals	—		—		19	8.0	19	4.6
P.G.H. Kisumu	5	8.3	10	8.7	13	5.5	28	6.8
Nyanza Prov.								
3 Distr. Hospitals	—		—		9	3.8	9	2.2
P.G.H. Kakamega	2	3.3	7	6.1	10	4.2	19	4.6
Western Prov.								
3 Distr. Hospitals	—		—		7	2.9	7	1.7
P.G.H. Garissa	2	3.3	3	2.6	—		5	1.2
N.E. Prov.								
2 Distr. Hospitals	—		—		1	0.4	1	0.2
Total	60	=100%	115	=100%	238	=100%	413	=100%
	=14.5%	—	27.8%	—	57.6%	—	413	=100%

Source: Ministry of Health, Nairobi, pers. comm. 1974)

Table XI. Relative frequency of 4 helminths

from stool examinations in 54 hospital laboratories in Kenya from "Annual Laboratory Returns" 1958 – 1973 (cumulative) grouped by hospitals with similar frequency distribution pattern (see Fig. 16 back auf Map 7)

Provinces	Cumulative No of Exam	Ascaris		Ankylost.		Taenia		S. mansoni	
		n	%	n	%	n	%	n	%
1. East Coast	145,963	21,679	31.2	33,496	65.8	1,656	2.2	451	0.5
2. Kamba	84,011	4,795	35.5	1,513	11.3	1,421	10.5	4,163	42.5
3. Aberdare	181,265	23,087	44.1	12,334	32.6	4,451	14.5	2,253	7.9
4. Nyeri-Mt. Kenya	267,298	39,221	72.4	10,338	14.3	5,225	11.4	1,030	1.7
5. Narok-NE	26,972	1,998	37.9	799	12.7	2,837	42.5	256	6.9
6. Rift Valley	96,078	8,109	37.7	3,776	14.2	9,692	47.6	144	0.5
7. Kisi	127,620	27,481	51.5	7,197	12.2	10,832	27.2	642	2.5
8. Nyanza	141,441	25,087	47.9	13,068	26.9	9,553	20.3	2,405	14.6
9. Western Province	29,916	2,561	27.1	5,968	51.1	25,199	18.4	101	0.3

Table XII. Schistosoma haematobium

findings from urine examinations in 54 hospital laboratories in Kenya from "Annual Laboratory Returns" 1958 – 1973, cumulative, grouped hospitals with similar frequency distribution pattern of intestinal helminths

Typ	Hospital	Urine examin.	pos n	pos %	Typ	Hospital	Urine examin.	pos n	pos %
1.	Mombasa	32,831	4,877	14.85	6.	Eldoret	15,942	74	0.46
	Msambweni	12,204	3,800	31.14		Kapsabet	6,549	56	0.86
	Kinango	10,657	3,499	32.83		Tambach	1,896	1	0.05
	Kilifi	39,917	5,802	14.54		Kitale	5,464	18	0.33
	Malindi	34,721	6,074	17.49		Kapenguria	283	0	0
	Kipini	2,606	653	40.66		Kabarnet	4,090	4	0.10
	Galole	20,133	9,103	45.21		Maralal	3,772	0	0
						Th. Falls	1,808	19	1.05
				28.10		Marsabit	—	—	—
									0.36
2.	Machakos	9,146	96	1.05					
	Kangundo	4,635	331	7.14	7.	Kisii	14,327	185	1.29
	Kitui	20,756	6,743	32.50		Kakamega	5,728	13	0.23
						Nandi Hills	1,605	67	4.17
				10.83		Kericho	13,166	162	1.23
						Londiani	2,283	67	2.93
3.	Kiambu	7,930	13	0.16		Nakuru	19,006	150	0.79
	Thika	3,284	50	1.52		Garissa	1,625	166	10.2
	Gatundu	—	—	—					
	Murang'a	16,929	32	0.19					1.67
	Kimbimbi	7,183	14	0.19					
	Kianyaga	2,335	5	0.21	8.	Kisumu	52,044	2,004	3.95
	Makindu	602	17	2.82		Homa Bay	18,155	2,113	11.66
	Voi	8,831	267	3.02					
	Taveta	8,515	3,886	45.64					7.75
				5.97	9.	Bungoma	6,312	15	0.24
						Alupe	1,032	1	0.10
4.	Nyeri	20,601	11	0.05		Busia	6,035	38	0.63
	Nanyuki	2,316	6	0.26					
	Naivasha	1,027	153	14.90					0.32
	Kerugoya	4,107	7	0.17					
	Baricho	—	—	—					
	Embu	10,972	255	2.32					
	Meru	—	—	—					
	Wesu	8,111	435	5.36					
	Lamu	10,160	748	7.36					
				3.38					
5.	Narok	8,199	32	0.39					
	Kajiado	4,312	7	0.16					
	Wajir	1,186	19	1.60					
	Mandera	1,726	69	4.0					
	Moyale	270	0	0					
				0.51					

Table XIII. Schistosomiasis: Summary of positive results of examinations of schoolchildren in some districts of Kenya

Province District Area	Author, year	No of children examined	S. mansoni		S. haematobium		max.	min.
			n	%	n	%	%	%
Coast Province								
Kwale, Shimba Hills	DVBD 1972	1,280	—	—	289	22.5	—	—
Kilifi	DVBD 1973	813	—	—	549	67.5	—	—
Eastern Province								
Kitui	DVBD 1972/74	5,812	1,020	17.5	—	—	26.5	2.5
Machakos town	Teesdale 1966	6,977	1,196	17.1	—	—		
Machakos Distr.	Mutinga a. Ngoka 1971	2,952	871	29.5	—	—	42.2	21.3
Central Province								
Kirinyaga	DVBD 1972/73	3,373	1,183	37.1	—	—	69.8	5.0
Murang'a	DVBD 1972/73	706	97	13.7	—	—	40.0	6.4
Kiambu	DVBD 1972/73	3,211	412	12.8	—	—	38.2	6.8
Nyanza Province								
Rusinga Isl.	Pamba 1974	916	272	30.0	—	—	60.0	7.0
Mfangano Isl.								
	Wijers, Munanga 1971	657	302	46.0	—	—	80.0	31.0
N. Nyakach	Kinoti 1971	774	—	—	187	37.7	58.3	17.5
Kano Plain	Kinoti 1971	1,708	—	—	60	3.5	5.0	0
Western Province								
Kakamega	DVBD							
Kimilili	1972	542	107	19.7	62	11.4	—	—

Tab. XIV. Gross domestic product by industrial origin 1964 and 1974

	At Constant (1964) Prices (K£ million)		Percentage of Total Gross Product at Constant (1964) Prices	
	1964	1974 *	1964	1974 *
Gross Product at Factor Cost				
A. Outside Monetary Economy				
Agriculture	73.47	104.18	22.3	17.2
Forestry	1.99	3.00	0.6	0.5
Fishing	0.11	0.13	—	—
Building and Construction	5.81	7.96	1.8	1.3
Water	2.09	2.88	0.6	0.5
Ownership of Dwellings	5.53	9.16	1.7	1.5
Total Product Outside Monetary Economy	89.00	127.31	27.0	21.0
B. Monetary Economy				
1. Enterprises and Non-Profit Institutions:				
Agriculture	53.08	89.22	16.1	14.7
Forestry	1.88	4.01	0.6	0.7
Fishing	0.85	1.04	0.3	0.2
Mining and Quarrying	1.46	3.42	0.4	0.6
Manufacturing and Repairing	34.17	76.11	10.4	12.5
Building and Construction	6.82	14.72	2.1	2.4
Electricity and Water	4.84	9.75	1.5	1.6
Transport, Storage and Communications	24.52	48.05	7.4	7.9
Wholesale and Retail Trade	32.98	47.01	10.0	7.7
Banking, Insurance, and Real Estate	9.85	23.56	3.0	3.9
Ownership of Dwellings	13.34	21.61	4.0	3.6
Other Services	11.90	34.16	3.6	5.6
Total	195.69	372.66	59.3	61.4
2. Private Households (Domestic Services)	2.94	4.78	0.9	0.8
3. General Government:				
Public Administration	16.84		5.1	
Defence	2.19		0.7	
Education	11.20		3.4	
Health	4.69		1.4	
Agricultural Services	4.41		1.3	
Other Services	3.13		0.9	
Total	42.47	102.47	12.8	16.9
Total Product Monetary Economy	241.10	479.91	73.0	79.0
Total Gross Product at Factor Cost (Monetary and Non-Monetary)	330.10	707.22	100.0	100.0
Gross Domestic Product Per Capita	36.26	47.03		

Source: Statistical Abstract 1975, Tab. 44 b, 45   * Provisional

References

Arap Siongok, T. K.; J. Birgen 1976: Cutaneous leishmaniasis: an occupational hazard. E. afr. med. J. 53, 292 – 294.

Bader, F. J. W. 1970: Die Vegetation Ostafrikas. Dargestellt am Kartenausschnitt der Serie Lake Victoria im Afrika-Kartenwerk der Deutschen Forschungsgemeinschaft. Berlin (mimeogr. unpublished).

Bagshawe, A. F.; H. M. Cameron 1974: Hepatitis B antigen and liver disease. In: Vogel, L. C., A. S. Muller, R. S. Odingo et al. (Edit.). Health and Diseases in Kenya, East African Literature Bureau, Nairobi, p. 285 – 292.

Baker de, W. C. 1963: A note on the epidemiology of tetanus in Kenya. E. afr. med. J. 40, 127 – 131.

Bakshi, A. K. 1974: Dental conditions and dental health. In: Vogel, L. C., A. S. Muller, R. S. Odingo et al. (Edit.). Health and Diseases in Kenya. East African Literature Bureau, Nairobi, p. 519 – 522.

Ball, M. G. 1966: Animal hosts of leptospires in Kenya and Uganda. Am. J. trop. Med. Hyg. 15, 523.

Barrows, H. H. 1923: Geography as human ecology. In: Ann. Ass. Am. Geogr. 13, 1 – 14.

Barwell, C. W. 1956: A note of some changes in the economy of the Kipsigis tribe. Journal of African Administration 8, 95 – 101.

Beck, A. 1970: A History of the British Medical Administration of East Africa 1900 – 1950. Cambridge, Mass. Harvard Univ. Press.

Bell, S. 1953: Predisposing factors in tick-borne relaping fever in Meru-district of Kenya. Trans. roy. Soc. trop. Med. Hyg. 47, 309 – 317.

Bell, S. 1956: The Ameru people of Kenya. J. trop. Med. Hyg., 59, 121 – 133.

Berry, L. 1969: Physical features. In: East Africa: its peoples and resources. Edited by W. T. W. Morgan. Nairobi, London, New York.

Bisley, G. G.; W. R. Burkitt 1974: Eye diseases. In: Vogel, L. C., A. S. Muller, R. S. Odingo et al. (Edit.). Health and Diseases in Kenya East African Literature Bureau Nairobi, p. 461 – 468.

Blanckenburg, P. v. und H. D. Cremer (Hrsg.) 1967: Handbuch der Landwirtschaft und Ernährung in den Entwicklungsländern. Bd. 1. Die Landwirtschaft in der wirtschaftlichen Entwicklung. Ernährungsverhältnisse. Stuttgart.

Blankhart, D. M. 1974: Human nutrition. In: Vogel, L. C., A. S. Muller, R. S. Odingo et al. (Edit.). Health and Diseases in Kenya. East African Literature Bureau Nairobi, p. 409 – 428.

Bohdale, M.; N. E. Gibbs, W. K. Simmons 1968: Incidence of endemic goitre in Kenya and outline for its prevention. Report to the Ministry of Health (mimeogr. unpublished), Nairobi.

Bohdale, M.; N. E. Gibbs, W. K. Simmons 1969: Nutrition survey and campaign against malnutrition in Kenya, 1964 – 1968. Mimeogr. report to the Ministy of Health (unpublished), Nairobi.

Bowen, E. T. W.; D. I. H. Simpson; G. S. Platt et al. 1973: Large scale irrigation and arbovirus epidemiology, Kano plain Kenya II. Preliminary serological Survey. Trans. roy. Soc. trop. Med. Hyg. 67, 702.

Brown, A. W. A. 1962: A Survey of simulium control in Africa (incl. Kenya, Uganda). Bull. Wld. Hlth. Org. 27, 511 – 527.

Brown, D. S. 1975: Distribution of intermediate hosts of Schistosoma on the Kano plain of Western Kenya. E. afr. med. J. 52, 42 – 51.

Brown, J. A. K. 1959: Factors influencing the transmission of leprosy. Trans. roy. Soc. trop. Med. Hyg. 53, 179 – 189.

Brown, L. H. 1968: Agricultural change in Kenya: 1945 – 1960. Food Research Institute Studies (Stanford University) 8, 33 – 90.

Buckley, J. J. C. 1949: Studies on human onchocerciasis and simulium in Nyanza Province, Kenya I. J. Helminth, 23, 1 – 24.

Buckley, J. J. C. 1950: Studies on human onchocerciasis and simulium in Nyanza Province, Kenya, II. J. Helminth, 25, 213 – 222.

Burdin, M. L.; G. Froyd; W. A. Ashford 1958: Leptospirosis in Kenya due to leptospira gripotyphosa. Vet. Rec. 70, 830.

Burkitt, D. P. 1951: Primary hydrocele and its treatment. Lancet 1, 1341 – 1348.

Burkitt, D. P.; Bundschuh; K. Dahlin; L. Dahlin; R. Neale 1969: Some cancer patterns in West-Kenya and North-West-Tanzania. E. afr. med. J. 46, 188 – 193.

Callcott, R. D. 1954: Blindness in Kenya. British Empire Society for the Blind. London.

Chowdhury, A. W. 1974: Seven years of snail control at Mwea Irrigation Settlement Kenya, Results and Costs. E. afr. Med. J. 51, 600 – 609.

Chowdhury, A. W. 1975: Potential effects of irrigation on the spread of Bilharziasis in Kenya. E. afr. med. J. 52, 120 – 126.

Colony and Protectorate of Kenya, Ministry of Health: Annual Report 1958, 1960, 1961. Nairobi 1959, 1961, 1962.

Cook, P.; D. Burkitt 1970: An epidemiological study of seven malignant tumours in East Africa. Medical Research Council. (mimeogr. unpub. Report). London.

Craddock, A. L. 1961: Some features of the epidemiology of tick typhus in Kenya. E. afr. med. J. 38, 228 – 231.

Craddock, A. L. 1974: Tick fever. In: Vogel, L. C., A. S. Muller, R. S. Odingo et al. (Edit.). Health and Diseases in Kenya. East African Literature Bureau, Nairobi, p. 249 – 252.

Craddock, A. L. 1974: Q-Fever. In: Vogel, L. C., A. S. Muller, R. S. Odingo et al. (Edit.). Health and Diseases in Kenya, East African Literature Bureau, Nairobi, p. 253 – 254.

Darragh, J. H.; A. R. Hutchinson; E. N. Mngola 1971: The diabetic clinic, Kenyatta National Hospital, review of results of treatment and recommendations. E. afr. med. J. 48, 327.

Davies, D. H. S.; R. B. Heisch; D. Mc Neill; K. F. Meyer 1968: Serological survey of plague in rodents and other small mammals in Kenya. Trans. roy. Soc. trop. Med. Hyg. 62, 838 – 861.

De Geus, A. 1974: Leptospirosis. In: Vogel, L. C., A. S. Muller, R. S. Odingo et al. (Edit.), Health and Diseases in Kenya, East African Literature Bureau, Nairobi, p. 245 – 248.

De Geus, A.; O. Kranendonk; H. J. Bohlander 1969: Clinical leptospirosis in Kwale District, Coast Province, Kenya. A pilot study. E. afr. med. J. 46, 491.

Diesfeld, H. J. 1965: Amöbiasis im tropischen Hochland. Z. Tropenmed. Parasit. 16, 403 – 410.

Diesfeld, H. J. 1969: Biostatistische Auswertung von Krankenhausberichten als Grundlage einer geomedizinischen Analyse der Krankheitsverbreitung in tropischen Entwicklungsländern (Habilitationsschrift). Heidelberg.

Diesfeld, H. J. 1969: Befundhäufigkeit von Helminthen beim Menschen in Kenia in Beziehung zu Umweltfaktoren. Z. Tropenmed. Parasit. 20, 310 – 333.

Diesfeld, H. J. 1969: Die geomedizinische Darstellung der regionalen Häufigkeitsverteilung von 5 Darmhelminthen in Kenia. Z. Tropenmed. Parasit. 20, 483 – 494.

Diesfeld, H. J. 1970 a: The evaluation of hospital returns in developing countries. Meth. Inform. Med., 9, 27 – 34.

Diesfeld, H. J. 1970 b: Beziehungen zwischen Häufigkeit des Hakenwurmvorkommens und des Klimas in Kenia, dargestellt am Temperatur-Feuchte-Milieu. Z. Tropenmed. Parasit. 21, 84 – 92.

Diesfeld, H. J. 1973: The definition of the hospital catchment area and its population as a denominator for the evaluation of hospital returns in developing countries. Int. J. epid. 2, 47 – 53.

Diesfeld, H. J. 1974: Zur Methodik der Darstellung der Raumbezogenheit von Krankheitsvorkommen. In: Fortschr. d. geomed. Forschung, hrsg. v. H. J. Jusatz, Beiheft 35 der Schriftenreihe "Erdkundliches Wissen" der Geogr. Zeitschrift, S. 126 – 141.

Diesfeld, H. J.; H. Hecklau 1976: Modell für eine medizinische Landeskunde, dargestellt am Beispiel Kenia. In: Methoden und Modelle der geomedizinischen Forschung, hrsg. v. H. J. Jusatz, Beiheft 43 der Schriftenreihe "Erdkundliches Wissen" der Geogr. Zeitschrift, S. 71 – 89.

Dissevelt, A. G.; L. C. Vogel 1971: An analysis of the operations of the medical assistant in an outpatient department with the emphasis on administrative procedures. In: Health and Disease in Africa. Proceedings of the East African Medical Research Council, Scientific Conference 1970. Editor: G. C. Gould, E. Afr. Lit. Bureau, Nairobi, p. 51 – 58.

Dissevelt, A. G. 1972: The influence of health centre activities on maternal and child health in a rural area in Kenya. Ann. Rep. Med. Res. Centre Nairobi, Roy. trop. Inst. Amsterdam, Nairobi, p. 93 – 100.

East African Dental Association 1973: East African Dental and Medical Directory, 1973. East African Literature Bureau Nairobi.

Emiru, V. P.; G. Dechet 1969: Trachoma Survey in Bukedi, Uganda. E. afr. med. J. 47, 30 – 37.

Engelhard, K. 1974: Die wirtschaftsräumliche Gliederung Ostafrikas. München. (Afrika-Studien Nr. 84).

Family planning in Kenya. 1967: A report submitted to the government of Kenya by an advisory mission of the Population Council of the United States of America. Published by the Ministry of Economic Planning and Development, Nairobi.

Fendall, N. R. E. 1952: Kala azar in East Africa with particular reference to Kenya and the Kamba Country. Part I – III. J. trop. Med. Hyg. 55, 193 – 204; 220 – 233; 245 – 256.

Fendall, N. R. E. 1960: Poliomyelitis in Kenya. E. afr. med. J. 37, 89 – 103.

Fendall, N. R. E. 1961: The spread of kala azar in Kenya. E. afr. med. J. 38, 417 – 419.

Fendall, N. R. E. 1962: Poliomyelitis in Kenya. The 1960 epidemic and oral vaccination campaign. J. trop. Med. Hyg. 65, 245 – 255.

Fendall, N. R. E. 1963: Health Centres, a basis for a rural health service. J. trop. Med. Hyg. 66, 219.

Fendall, N. R. E.; J. G. Grounds. 1965: The incidence and epidemiology of disease in Kenya. Part I. Some diseases of social significance. J. trop. Med. Hyg. 68, 77 – 84.

Fendall, N. R. E.; J. G. Grounds 1965 b: The incidence and epidemiology of disease in Kenya. Part II. Some important communicable diseases. J. trop. Med. Hyg. 68, 113 – 120.

Fendall, N. R. E.; J. G. Grounds. 1965 c: The incidence and epidemiology of disease in Kenya. Part III. Insect borne diseases. J. trop. Med. Hyg. 68, 134.

Fendall, N. R. E.; B. M. Lake 1958: Poliomyelitis in Kenya. J. trop. Med. Hyg. 61, 135.

Fliedner, H. 1965: Die Bodenrechtsreform in Kenya. Studie über die Änderung der Bodenrechtsverhältnisse im Zuge der Agrarreform unter besonderer Berücksichtigung des Kikuyu-Stammesgebietes. Berlin, Heidelberg, New York. (Afrika-Studien Nr. 7).

Fliedner, H. 1968: Die Wandlung der Agrarstruktur in Kenia. Geographische Rundschau 20, 81 – 86.

Forrester, A. T. T.; O. Kranendonk; L. H. Turner; T. W. Wolff; H. J. Bohlander 1961: Serological evidence of human leptospirosis in Kenya. E. afr. med. J. 46, 497 – 506.

Forrester, A. T. T.; G. S. Nelson; G. Sander 1961: The first outbreak of trichinosis in Africa south of the Sahara. Trans. roy. Soc. trop. Med. Hyg. 55, 503 – 513.

Foy, H.; A. G. Kendall 1974: Haemoglobinopathies. In: Vogel, L. C., A. S. Muller, R. S. Odingo et al. (Edit.), Health and Diseases in Kenya, East African Literature Bureau Nairobi, p. 437 – 444.

Foy, H.; A. Kondi 1960: The relation of hookworm load and species to intestinal blood loss and the genesis of iron deficiency anaemia. Trans. roy. Soc. trop. Med. Hyg. 55, 26 – 29.

Garnham, P. C. C.; C. W. Davies; R. B. Heisch; G. L. Timms 1947: An epidemic of louseborne relapsing fever in Kenya. Trans. roy. Soc. trop. Med. Hyg. 41, 141 – 170.

Geser, A.; S. Christensen; I. B. Thorup 1970: A multipurpose serological survey in Kenya. I. Survey methods and program of field work. Bull. Wld. Hlth. Org. 43, 521 – 537.

Golskowsky, R. 1969: Bewässerungslandwirtschaft in Kenya. Darstellung grundsätzlicher Zusammenhänge am Beispiel des Mwea Irrigation Settlement. München. (Afrika-Studien. 39).

Griffiths, J. F. 1969: Climate. In: East Africa: its peoples and resources. Edited by W. T. W. Morgan. Nairobi, London, New York.

Haddow, A. J. A. 1952: A review of the results of yellow fever protection tests on the sera of primates in Kenya. Ann. trop. Med. Parasit. 46, 135.

Haddow, A. J.; C. W. Davies; A. J. Walker 1960: O'Nyong nyong fever: an epidemic virus disease in East Africa. 1. Introduction Trans. roy. Soc. trop. Med. Hyg. 54, 517 – 522.

Halliman, D. M.; W. T. W. Morgan 1967: The City of Nairobi. In: Nairobi: City and Region. Edited by W. T. W. Morgan. Nairobi, London, New York. pp. 98 – 120.

Hanegraaf, Th. A. C. 1974: Thyroid Diseases. Population Based Studies of Endemic Goitre. In: Vogel, L. C., A. S. Muller, R. S. Odingo et al. (Edit.), Health and Diseases in Kenya, East African Literature Bureau Nairobi, p. 395 – 400.

Hanegraaf, Th. A. C.; P. E. McGill 1970: Prevalence and geographical. distribution of endemic goitre in East Africa. E. afr. med. J., 47, 61 – 65.

Hanegraaf, Th. A. C.; P. E. McGill; J. R. Taylor 1971: Endemic goitre and its prevention in Kenya. In: Health and Disease in Africa. p. 60 – 67. Ed.: C. Gould. Nairobi, East African Literature Bureau.

Hartmann, A. 1973: The prevalence of leprosy at the coast of Kenya. E. afr. med. J., 50, 181 – 188.

Haynes, W. S. 1951: Tuberculosis in Kenya. Colony & Protectorate of Kenya. Govt. Printer, Nairobi

Hecklau, H. 1967: Landwirtschaftliche Flächennutzungsstile im Gebiet des Kartenblattes Lake Victoria. Die Erde 98, 135 – 142. (Vorläufige Ergebnisse der Untersuchungen im Rahmen des Afrika-Kartenwerkes der Deutschen Forschungsgemeinschaft. 1.).

Hecklau, H. 1968: Die agrarlandschaftlichen Auswirkungen der Bodenbesitzreform in den ehemaligen White Highlands von Kenya. Die Erde 99, 236 – 264.

Hecklau, H. 1969: Das Uasin-Gishu-Trans-Nzoia-Plateau im Hochland von Kenya. Eine wirtschafts- und bevölkerungsgeographische Skizze. Ostafrikanische Studien. Ernst Weigt zum 60. Geburtstag. Nürnberger wirtschafts- und sozialgeographische Arbeiten. 8, 168 to 191. Nürnberg.

Hecklau, H. 1970: Bewässerungsfeldbau in Kenya. In: Boesler, K.-A. u. Kühn, A. (ed.): Aktuelle Probleme geographischer Forschung. Festschrift für Joachim Heinrich Schultze aus Anlaß seines 65. Geburtstages. Berlin, 475 – 492. (Abhandlungen des Geographischen Instituts der Freien Universität Berlin. 13.).

Hecklau, H. 1974: Irrigation Farming in Kenya. In: Applied Sciences and Development. A Biannual Collection of Recent German Contributions Concerning Development through Applied Sciences. 4, 75 – 88.

Hecklau, H. 1976: Landwirtschaftliche Flächennutzungsstile/Modes of Agricultural Land Use/Modes de mise en valeur des surfaces agricoles. Map 1 : 1 Mio. In: Afrika-Kartenwerk. Edited on behalf of the German Research Society by K. Kayser, W. Manshard, H. Mensching, J. H. Schultze.

Hecklau, H. 1978: Agargeographie im äquatorialen Ostafrika. Landwirtschaftliche Flächennutzungsstile in Ostafrika im Spannungsfeld zwischen Tradition und Fortschritt. Afrika-Kartenwerk. Serie E. Beiheft zu Blatt 12 (with English and French Summary). Edited on behalf of the German Research Society by E. Kayser, W. Manshard, H. Mensching, J. H. Schultze.

Henderson, B. E. 1968: Trachoma in Uganda. In: Uganda Atlas of Disease Distribution. Kampala.

Heisch, R. B. 1954: Studies in leishmaniasis in East Africa. I. The epidemiology of an outbreak of kala azar in Kenya. Trans. roy. Soc. trop. Med. Hyg. 48, 449 – 464.

Heisch, R. B. 1960: The isolation of Rickettsia burneti from lemnis-comys Sp. in Kenya. E. afr. med. J. 37, 104.

Heisch, R. B. 1961: Rodents as reservoir of anthropod borne diseases in Kenya. E. afr. med. J. 38, 256 – 261.

Heisch, R. B.; C. A. W. Guggisberg; C. Teesdale 1956: Studies in leishmaniasis in East Africa. II. The sandfly of the Kituikala azar area in Kenya with description of 6 new species. Trans. roy. Soc. trop. Med. Hyg. 50, 209 – 226.

Heisch, R. B.; W. E. Grainger; A. E. C. Harvey; G. Lister 1962: Feral aspects of rickettsial infections in Kenya. Trans. roy. Soc. trop. Med. Hyg. 56, 272 – 286.

Heisch, R. B.; W. E. Grainger; I. S. A. M. D'Souza 1953: Results of plague investigations in Kenya. Trans. roy. Soc. trop. Med. Hyg. 47, 503 – 521.

Heisch, R. B.; J. P. McMahon; P. E. C. Manson-Bahr 1958: The isolation of Trypanosoma rhodesiense from a bushbuck. Brit. med. J. 2, 1203.

Heisch, R. B.; R. McPhee; L. R. Rickmann 1957: The epidemiology of tick typhus in Nairobi. E. afr. med. J. 34, 459 – 477.

Herskovits, M. J. 1926: The cattle complex in East Africa. American Anthropologist 28, 230 – 272, 361 – 388, 494 – 528, 633 – 644.

Heyer, J.; D. Ireri; J. Moris 1971: Rural development in Kenya. Nairobi, Dar es Salaam, Kampala. (East African Rural Development Studies. 6).

Heyer, J.; J. K. Maitha; W. M. Senga 1976: Agricultural development in Kenya. An Economic Assessment. Nairobi, Lusaka, Dar es Salaam, Addis Ababa.

Highton, R. B. 1974 a: Health risks in water conservation schemes. In: Vogel, L. C.; A. S. Muller; R. S. Odingo et al. (Edit.)., Health and Diseases in Kenya, East African Literature Bureau Nairobi, p. 175 – 180.

Highton, R. B. 1974 b: Schistosomiasis. In: Vogel, L. C., A. S. Muller, R. S. Odingo et al. (Edit.), Health and Diseases in Kenya. East African Literature Bureau, Nairobi, p. 347 – 356.

Highton, R. B. 1974 c: Onchocerciasis. In: Vogel, L. C., A. S. Muller, R. S. Odingo et al. (Edit.), Health and Diseases in Kenya, East African Literature Bureau, Nairobi, p. 363 – 368.

Huntingford, G. W. B. 1950: Nandi work and culture. London.

Huntingford, G. W. B. 1953: The Nandi of Kenya: Tribal. control in a pastoral society. London.

Huntingford, G. W. B. 1953: The southern Nilo-Hamites. Ethnographic survey of Africa. East Central Africa. Part VIII. Herausgegeben von D. Forde. London.

Illies, J. H. and St. Mueller (Ed.) 1970: Graben Problems. Proceedings of an International Rift Symposium held in Karlsruhe October, 10 – 12, 1968. Stuttgart.

Innes, J. R. 1949: Leprosy in Kenya. E. afr. med. J. 26, 32 – 35.

International Bank for Reconstruction and Development 1963: The economic development of Kenya. Report of a mission organized by the International Bank for Reconstruction and Development at the request of the governments of Kenya and the United Kingdom. Baltimore.

Jätzold, R. 1967: Aktuelle Probleme der Europäersiedlungen in Ostafrika. Geographische Zeitschrift 55, 42 – 51.

Jätzold, R. 1969: Die Feindifferenzierung des Klimas der Tropen durch eine Dezimalklassifikation. Tübinger Geogr. Studien. Wilhelmy-Festschrift.

Jätzold, R. 1970: Ein Beitrag zur Klassifikation des Agrarklimas der Tropen (mit Beispielen aus Ostafrika). In: Tübinger Geogr. Studien 34, 57 – 69. Beiträge zur Geographie der Tropen und Subtropen. Festschrift für Herbert Wilhelmy. Tübingen. (Tübinger Geogr. Studien Sonderband 3).

James, L. 1939: The Kenya Masai, a nomadic people under modern administration. Africa 12, 49 – 73.

Jusatz, H. J. 1961: Cerebrospinalmeningitis. In: Welt-Seuchen-Atlas – World Atlas of Epidemic Diseases – Bd. III, 39 – 44, hrsg. v. E. Rodenwaldt u. H. J. Jusatz, Hamburg.

Jusatz, H. J. 1963: Richtlinien für die Abfassung von Medizinischen Länderkunden. Arch. Hyg. Bakt. 147, 280 – 288.

Jusatz, H. J. 1974: Geomedizinische Grundlagen für eine Geoökologie der Infektionskrankheiten. In: Fortschr. d. geomed. Forschung, hrsg. v. H. J. Jusatz, Beiheft 43 der Schriftenreihe "Erdkundliches Wissen" der Geogr. Zeitschrift, S. 1 – 29.

Jusatz, H. J. 1976: Zielvorstellungen geomedizinischer Forschung. In: Methoden und Modelle geomedizinischer Forschung, hrsg. v. H. J. Jusatz. Beiheft 43 der Schriftenreihe "Erdkundliches Wissen", S. 1 – 12.

Kent, P. W. 1974: Tuberculosis. In: Vogel, L. C., A. S. Muller, R. S. Odingo et al. (Edit.), Health and Diseases in Kenya, East African Literature Bureau, Nairobi, 193 – 204.

Kenyatta, J. 1953: Facing Mount Kenya. The Tribal Life of the Gikuyu. (Reprinted, 1st. Ed. 1938) London.

King, M. 1966: Medical Care in Developing Countries. Oxford Univ. Press London, Nairobi, Lusaka, Addis Abeba.

Kinoti, G. K. 1971: The epidemiology of Schistosoma haematobium infection on the Kano Plain of Kenya. Trans. roy. Soc. trop. Med. Hyg. 65, 637 – 645.

Kinoti, G. K. 1971: Epidemiology of Schistosoma mansoni infection on the Kano Plain of Kenya. Trans. roy. Soc. trop. Med. Hyg. 65, 646 – 649.

Koinange Karuga, W.; J. J. Rogowski, D. Metselaar 1973: Poliomyelitis: epidemiology and prophylaxis, 3. Nationwide vaccination campaign against poliomyelitis in Kenya with the help of laymen. Bull. Wld. Hlth Org. 48, 543 – 545.

Koinange Karuga, W. 1974: Smallpox. In: Vogel, L. C., A. S. Muller, R. S. Odingo et al. (edit.), Health and Diseases in Kenya, East African Literature Bureau, Nairobi, 261 – 266.

Kranendonk, O.; J. W. Wolff; H. J. Bohlander; J. M. D. Roberts; A. DeGeus; R. Njenga 1967/68: Leptospirosis. Ann. Rep. Med. Res. Centre, Nairobi, Roy. trop. Inst. Amsterdam.

Langlands, B. W. 1967: Sleeping Sickness in Uganda 1908 – 1920. Occasional Papers No. 1. Dept. of Geography Makerere University College, Kampala, Uganda.

Leiker, D. L. 1966: Leprosy in Kenya (unpublished report). Medical Research Centre, Nairobi, Roy. trop. Inst. Amsterdam.

Linsell, C. A. 1974: Cancer. In: Vogel, L. C., A. S. Muller, R. S. Odingo et al. (Edit.), Health and Diseases in Kenya, East African Literature Bureau, Nairobi, 381 – 392.

Luijk, J. N. van 1974: Social and Cultural Aspects of Health and Disease. In: Vogel, L. C., A. S. Muller, R. S. Odingo et al. (Edit.), Health and Disease in Kenya, East Africa Literature Bureau, Nairobi, 63 – 74.

Lurtz, R. 1913: Eine Pestepidemie am Kilimanjaro 1912. Arch. Schiffs-u. Tropenhyg. 17, 593 – 599.

Mahaffy, A. F.; K. C. Smithburn; T. P. Hughes 1946: Immunity to yellow fever. Trans. roy. Soc. trop. Med. Hyg. 40, 57 – 82.

Manshard, W. 1968: Agrargeographie der Tropen. Eine Einführung. Hochschultaschenbücher 356/356 a. Mannheim/Zürich.

Matson, A. T. 1957: The history of malaria in Nairobi. E. afr. med. J. 34, 431 – 441.

McCrae, A. W. R. 1968: Malaria in Uganda. In: Uganda Atlas of Disease Distribution, Kampala.

McKinnon, J. A. 1962: Kala azar in the Upper Rift Valley of Kenya. Part I. Background and discovery of the Disease. J. trop. Med. Hyg. 65, 51 – 63.

McKinnon, J. A. 1962: Kala azar in the Upper Rift Valley of Kenya. Part II. epidemiol. factors. J. trop. Med. Hyg. 65, 82 – 90.

McKinnon, J. A.; N. R. E. Fendall 1956: Kala azar in the Baringo District of Kenya. J. trop. Med. Hyg. 59, 208 – 212.

McMahon, J. P.; R. B. Highton; H. Goiny 1958: The eradication of Simulium neavei from Kenya. Bull. Wld. Hlth. Org. 19, 75 – 107.

Mello, D. J. P. 1947: Some aspects of malaria in Kenya. E. afr. med. J. 24, 112 – 126.

Merrill, R. S. 1960: "Resistance"to economic change: the Masai. Proceedings of the Minnesota Academy of Science 28, 120 – 131.

Metselaar, D. 1974: Arthropod-borne viral disease. In: Vogel, L. C., A. S. Muller, R. S. Odingo et al. (Edit.), Health and Diseases in Kenya, East African Literature Bureau, Nairobi, 273 – 278.

Metselaar, D. 1974: Yellow fever. In: Vogel, L. C., A. S. Muller, R. S. Odingo et al. (Edit.), Health and Diseases in Kenya, East African Literature Bureau, Nairobi, 279 – 284.

Metselaar, D.; Baldev Kaur Nottay 1974: Poliomyelitis. In: Vogel, L. C., A. S. Muller, R. S. Odingo et al. (Edit.), Health and Diseases in Kenya, East African Literature Bureau, Nairobi, 255 – 260.

Metselaar, D.; S. K. Dola; W. Gemert 1973: Poliomyelitis: epidemiology and prophylaxis, 2. Distribution of oral trivalent vaccine by laymen—volunteers. Bull. Wrld. Hlth. Org. 48, 429 – 433.

Metselaar, D.; B. E. Henderson; G. B. Kirya; P. M. Tukei; A. deGeus 1974: Isolation of arboviruses in Kenya, 1966 – 1971, Trans. roy. Soc. trop. med. 68, 114 – 123.

Metselaar, D.; B. E. Henderson; E. Kirya; G. L. Timms 1970: Recent research on yellow fever in Kenya. E. afr. med. J. 47, 1.

Middleton, J.; G. Kershaw 1965: The central tribes of the northeastern Bantu. (The Kikuyu, including Embu, Meru, Mbere, Chuka, Mwimbi, Tharaka, and the Kamba of Kenya). London (Ethnographic Survey of Africa. East Central Africa. Part 5).

Migue, M. R. 1974: Paramedical education. In: Vogel, L. C., A. S. Muller, R. S. Odingo et al. (Edit.), Health and Diseases in Kenya, East African Literature Bureau, Nairobi, 147 – 154.

Mingola, E. N. 1974: Cholera. In: Vogel, L. C., A. S. Muller, R. S. Odingo et al. (Edit.), Health and Diseases in Kenya, East African Literature Bureau, Nairobi, 181 – 184.

Mingola, E. N. 1974: Diabetes Mellitus. In: Vogel, L. C., A. S. Muller, R. S. Odingo et al. (Edit.), Health and Diseases in Kenya, East African Literature Bureau, Nairobi, 405 – 408.

Morley, D. 1973: Pediatric priorities in developing countries. London, p. 217.

Molnos, A. 1968: Attitudes towards family planning in East Africa. An investigation in schools around Lake Victoria and in Nairobi. With introductory chapters on the position of women and the population problem in East Africa. Afrika-Studien 26, München.

Morgan, W. T. W. 1963: The 'White Highlands' of Kenya. Geographical Journal 129, 140 – 155.

Morgan, W. T. W. 1965: Density of population map 1962. Kenya 1 : 1 Mio. Nairobi.

Morgan, W. T. W. 1966: Kenya. Population distribution, 1962. 1 : 1 Mio. Nairobi.

Morgan, W. T. W. (Ed.) 1967: Nairobi: City and Region. Nairobi, London, New York.

Morgan, W. T. W. (Ed.) 1969: East Africa: its peoples and resources. Nairobi. London, New York.

Morgan, W. T. W. 1973: East Africa. Geographies for advanced studies. Ed. S. H. Beaver. London.

Morgan, W. T. W.; N. M. Shaffer 1966: Population of Kenya. Density and distribution. A geographical introduction to the Kenya population census, 1962. Nairobi, Lusaka, Addis Abeba.

Msangi, A. S. 1969: Entomological observations after the 1968 plague outbreak in Mbulu District Tanzania. E. afr. med. J. 46, 465 – 470.

Muller, A. S. 1974: Tetanus. In: Vogel, L. C., A. S. Muller, R. S. Odingo et al. (Edit.), Health and Diseases in Kenya, East African Literature Bureau, Nairobi, 241 – 244.

Musiga, L. O. 1974: Problems of social protection in Kenya. Int. Soc. Sec. Review 4, 12 – 12.

Mustafa, G. 1974: Mental health and mental disorders. In: Vogel, L. C., A. S. Muller, R. S. Odingo et al. (Edit.), Health and Diseases in Kenya, East African Literature Bureau, Nairobi, 453 – 460.

Mutinga, M. J. 1975: The animal reservoir of cutaneous leishmaniasis on Mount Elgon, Kenya. E. afr. med. J. 52, 142 – 151.

Mutinga, M. J. 1975: Phlebotomus fauna in the cutaneous leishmaniasis of Mount Elgon, Kenya. E. afr. med. J. 52, 340 – 347.

Mutinga, M. J.; J. M. Ngoka 1971: Prevalence of intestinal schistosomiasis in Machakos District, Kenya. E. afr. med. J. 48, 559 – 563.

Mutinga, M. J.; J. M. Ngoka 1975: Cutaneous leishmaniasis in Kenya: montenegro skin test in leishmaniasis foci and presumably leishmaniasis-free areas. E. afr. med. J. 52, 333 – 339.

National Atlas of Kenya. 3rd Ed. Drawn printed and published by Survey of Kenya. Nairobi 1970.

Nelson, G. S.; J. G. Grounds 1958: Onchocerciasis at Kodera (Nyanza P.) 11 years after eradication of the vector. E. afr. med. J. 35, 365 – 368.

Nelson, G. S.; E. J. Blackie; J. M. Mukundi 1966: Comparative studies on geographical strains of Trichinella spiralis. Trans. roy. Soc. trop. Med. Hyg. 60, 471.

Nelson, G. S.; C. W. A. Guggisberg; J. Mukundi 1963: Animal host of Trichinella spiralis in East Africa. Ann. trop. Med. Parasit. 57, 332.

Nelson, G. S.; R. B. Heisch; M. Furlong 1962: Studies in filariasis in East Africa. II. Filarial infections in man, animals and mosquitos on the Kenyan coast. Trans. roy. Soc. trop. Med. Hyg. 56, 202 to 217.

Nelson, G. S.; R. L. Rausch 1963: Echinococcus infections in man and animals in Kenya. Ann. trop. Med. Parasit. 57, 136 – 149.

Nottay Kaur, B.; D. Metselaar 1973: Poliomyelitis, epidemiology and prophylaxis: 1. A. Longitudinal epidemiological survey in Kenya. Bull. Wld Hlth Org. 48, 421 – 427.

Nottidge, C. P. R.; J. R. Goldsack 1965: The Million—acre settlement scheme 1962 – 1966. Nairobi.

O'Connor, A. M. 1971: An economic geography of East Africa. 2nd. Ed. London.

Odingo, R. S. 1969: Observations on land use and settlement in the Kenya Highlands. In: Ostafrikanische Studien. Ernst Weigt zum 60. Geburtstag. Nürnberger wirtschafts- und sozialgeographische Arbeiten. 8. S. 254 – 278. Nürnberg.

Odingo, R. S. 1971: The Kenya Highlands. Land use and agricultural development. Nairobi. (African Geographical Studies 1.).

Ogendo, R. V. 1972: Industrial geography of Kenya. With special emphasis on the agricultural processing and fabricating industries. Nairobi. (African Geographical Studies 2.).

Ojany, F. 1970: Kenya Physical. In: National Atlas of Kenya. 3 rd. Edition, p. 4.

Ojany, F.; R. B. Ogendu 1973: Kenya. A Study in Physical and Human Geography. London.

O'Leary, P. 1976: A Five-year review of human hydatid cyst disease in Turkana District, Kenya. E. afr. Med. J. 53, 540 – 544.

Ominde, S. H. 1963: Problems of land and population in the Lake districts of Western Kenya. Reprint from first Proceedings East African Academy, Makerere. 1963.

Ominde, S. H. 1965: The ethnic map of the Republic of Kenya. Occasional Memoir No. 1. Department of Geography. University College, Nairobi.

Ominde, S. H. 1968: Land and population movements in Kenya. London, Nairobi, Ibadan.

Ominde, S. H. 1969: Internal migration of the economical active age group in Kenya. In: Ostafrikanische Studien. Ernst Weigt zum 60. Geburtstag. Nürnberger Wirtschafts- und Sozialgeographische Arbeiten. 8. S. 227 – 240. Nürnberg.

Ominde, S. H. (Ed.) 1971: Studies in East African geography and development. London, Nairobi, Ibadan.

Onyango, R. J.; van Hoeve; K.; P. de Raadt 1966: The epidemiology of Trypanosoma rhodesiense sleeping sickness in Alego Location, Central Nyanza, Kenya. Trans. roy. Soc. trop. Med. Hyg. 60, 175.

Onyango, R. J. 1974: African human trypanosomiasis (Sleeping Sickness). In: Vogel, L. C., A. S. Muller, R. S. Odingo et al. (Edit.), Health and Diseases in Kenya, East African Literature Bureau, p. 319 – 330, Nairobi.

Onyango, Z. 1974: Health facilities and services in Kenya. In: Vogel, L. C., A. S. Muller, R. S. Odingo et al. (Edit.), Health and Diseases in Kenya. East Africa Literature Bureau, p. 107 – 126. Nairobi.

Oomen, L. J. A.; J. Wegener 1974: Brucellosis. In: Vogel, L. C., A. S. Muller, R. S. Odingo et al. (Edit.), Health and Diseases in Kenya, East African Literature Bureau. Nairobi, p. 221 – 224.

Oomen, L. J. A. 1976: Human brucellosis in Kenya. Trop. Geogr. Med. 28, 45 – 53.

Ostafrikanische Studien. Ernst Weigt zum 60. Geburtstag. Nürnberger wirtschafts- und sozialgeographische Arbeiten. 8. 1968. Nürnberg.

Ouma, J. P. M. B. 1970: Evolution of tourism in East Africa (1900 to 2000). Nairobi, Dar es Salaam, Kampala.

Pamba, H. O. 1974: Schistosomiasis in Nyanza, Province, Kenya, Rusinga Island. E. afr. med. J. 51, 594 – 599.

Piercy, S. E. 1951: Canine leptospirosis in Kenya. Vet. Rec. 63, 425.

Pollitzer, R. 1954: Plague. World Health Org. Monogr. Ser. No. 22, Geneve.

Pratt, D. J.; P. J. Greenway; M. D. Gwynne 1966: A classification of East African rangeland, with an appendix on terminology. Journal of applied ecology 3, 369 – 382.

Radojcic, V. 1973: Cholera el tor vibrios and atypical formes isolated during the outbreak of cholera in Kenya in 1971. E. afr. med. J. 50, 696 – 704.

Raval, S. K. 1974: Traffic accidents and casualties. In: Health and Diseases in Kenya. Ed. by Vogel, L. C., A. S. Muller and R. S. Odingo, Z. Onyango, A. De Geus. Nairobi, p. 523 – 529.

Rees, P. H.; E. N. Mngola; Patricia O'Leary and H. O. Pamba 1974: intestinal parasites. In: Vogel, L. C., A. S. Muller, R. S. Odingo et al. (Edit.), Health and Diseases in Kenya, East African Literature Bureau, p. 339 – 346, Nairobi.

Republic of Kenya 1969: African socialism and its application to planning Kenya. Nairobi.

Republic of Kenya 1969: The constitution of Kenya. Kenya Gazette, Suppl. No. 27 (Acts No. 3), 18. April 1969.

Republic of Kenya 1965: Development plan 1966 – 1970. Nairobi.

Republic of Kenya 1969: Development plan 1970 – 1974. Nairobi.

Republic of Kenya 1974: Development plan 1974 – 1978. Nairobi.

Republic of Kenya: Economic survey 1973. Central Bureau of Statistics, Ministry of Finance and Planning. Nairobi. 1973.

Republic of Kenya: Estimates of recurrent expenditure, 1963/64, 1968/69, 1971/72, 1973/74. Nairobi 1963, 1968, 1971, 1973.

Republic of Kenya: Kenya population census, 1962. Vol. 1 – 4, Nairobi. 1.2. Tables. Advanced report. 1964. 3. African population. 1966. 4. Non-African population. 1966.

Republic of Kenya: Kenya population census, 1969. Vol. 1 – 3, Nairobi 1970 – 1971.

Republic of Kenya 1970: Laws of Kenya. Public Health Act, 1963, Revisions 1964, 1970. Nairobi.

Republic of Kenya 1975: Statistical abstract. Nairobi.

Republic of Kenya, Department of Veterinary Services. Annual report 1962/63. Nairobi, 1965, 1966.

Republic of Kenya, Ministry of Economic Planning and Development. Statistic Division: Economic survey 1959, 1961, 1962. Nairobi.

Republic of Kenya, Ministry of Health: Annual report, 1964, 1965, 1966. Nairobi, 1967, 1969, 1971.

Republic of Kenya, Ministry of Health and Housing: Annual report 1962, 1963. Nairobi, 1963, 1966.

Republic of Kenya, Ministry of Health 1968: The closer integration of the Health Services in Kenya. Report of a subcommittee appointed by the Central Advisory Board on Medical Missions. Nairobi.

Republic of Kenya, Ministry of Health 1972: Proposal for the improvement of Rural Health Services and the development of Rural Health Training Centres in Kenya. Nairobi.

Republic of Kenya, Ministry of Health. Division of vectorborne disease: Annual report 1965, 1969, 1970, 1971, 1972, 1973. Nairobi.

Republic of Kenya, Ministry of Health. Division of vectorborne disease: Annual report, 1973, Mombasa.

Republic of Kenya, Ministry of Health 1967: Inventory of health facilities in Kenya. Nairobi (mimeographed).

Rigby, E. P. 1960: The increasing importance of tetanus in Kenya. E. afr. med. J. 37, 1 – 21.

Roberts, J. I. 1950: The transmission of plague in Kenya. J. trop. Med. Hyg. 53, 80 – 87 u. 103 – 109.

Roberts, J. M. D. 1956: Pyrimethamine (Daraprim) in the control of epidemic malaria. J. trop. Med. Hyg. 59, 201.

Roberts, J. M. D. 1964: The control of epidemic malaria in the Highlands of Western Kenya. I. Before the campaign. J. trop. Med. Hyg. 67, 161 – 168.

Roberts, J. M. D. 1974: Malaria. In: Vogel, L. C., A. S. Muller, R. S. Odingo et al. (Edit.), Health and Diseases in Kenya, East African Literature Bureau, p. 305 – 318, Nairobi.

Roberts, J. M. D.; E. Neumann; C. W. Göckel; R. B. Highton 1967: Onchocerciasis in Kenya, 9, 11, 18 years after elimination of the vector. Bull. Wld Hlth Org. 37, 195 – 212.

Robinson, M. C. 1955: An epidemic of virus disease in the southern province, Tanganyika territory part I. Trans. roy. Soc. trop. Med. Hyg. 49, 28.

Roelsgaard, E.; J. Nyboe 1961: A tuberculosis survey in Kenya. Bull. Wld. Hlth. Org. 25, 851 – 870.

Rothermund, I. 1966: Die asiatische Minderheit in Ostafrika. Internationales Afrika Forum. 2, S. 411 – 414.

Ruthenberg, H. 1966: African agricultural production development policy in Kenya 1952 – 1965. Berlin, Heidelberg, New York (Afrika-Studien. 10.).

Ruthenberg, H. 1966: Der wirtschaftliche Erfolg landwirtschaftlicher Förderungsmaßnahmen in Kenya. Afrika heute. 9, S. 194 – 199.

Ruthenberg, H. 1967: Organisationsformen der Bodennutzung und Viehhaltung in den Tropen und Subtropen, dargestellt an ausgewählten Beispielen. In: Blanckenburg, P. V. u. H. D. Cremer (ed).: Handbuch der Landwirtschaft und Ernährung in den Entwicklungsländern. Stuttgart. Bd. 1. S. 122 – 208.

Say, P. J.; J. N. Itotia; I. A. Wamola 1974: Meningitis (Bacterial and Fungal). In: Vogel, L. C., A. S. Muller, R. S. Odingo et al. (Edit.), Health and Diseases in Kenya, In: East African Literature Bureau, p. 237 – 240, Nairobi.

Schneider, H. K. 1957: The subsistence role of cattle among the Pokot and in East Africa. American Anthropologist. 59, 278 – 300.

Schneider, H. K. 1959: Pokot resistance to change. In: Continuity and change in African cultures. Hrsg. von W. R. Bascom & M. J. Herskovits. Chicago. S. 144 – 167.

Schram, R. 1968: Road Accidents. In: Uganda Atlas of Disease Distribution. Makerere University College, Kampala. p. 166.

Schultze, J. H. 1966: Evolution and Revolution in der Landschaftsentwicklung Ostafrikas. Beiheft 14 der Schriftenreihe "Erdkundliches Wissen", der Geogr. Zeitschrift, Wiesbaden.

Schultze, J. H.; H. Hecklau; F. J. W. Bader 1967: Vorläufige Ergebnisse der Untersuchungen im Rahmen des Afrika-Kartenwerkes der Deutschen Forschungsgemeinschaft. Die Erde, 98, 135 – 149.

Seymour-Price, M.; C. Cachia; N. R. E. Fendall 1960: Smallpox in Kenya. E. afr. med. J. 37, 670 – 675.

Sorre, M. 1947: Les fondements biologiques de la Géographie humaine. Essai d'une écologie de l'homme. Paris, 2. Aufl. 1947.

Sorrenson, M. P. K. 1967: Landreform in the Kikuyu country. A study in government policy. Published on behalf of the East African Institute of Social Research. Nairobi, London.

Sorrenson, M. P. K. 1968: The origin of European settlement in Kenya. London.

Southgate, B. A.; B. V. E. Oriedo: Studies in the epidemiology of East African leishmaniasis. Trans. roy. Soc. trop. Med. Hyg. 56, 30 – 42.

Spencer, J. 1962: Some clinical aspects of bankroftian filariasis in the Lango District of Uganda. J. trop. Med. Hyg. 65, 256 – 261.

Stones, R. Y. 1939: Trachoma in Uganda day schools. E. afr. med. J. 16, 220 – 227.

Sturrock, R. F. 1965: The development of irrigation and its influence on the transmission of bilharziasis in Tanganyika. Bull. Wld Hlth Org. 32, 225 – 236.

Swynnerton, R. J. M. 1954: A plan to intensify the development of African agriculture in Kenya. Nairobi.

Swynnerton, R. J. M.; J. E. P. Booth; J. T. Moon: Report on agrian policy dealing with population increase, land tenure and fragmentation in Kenya. Nairobi, o. J.

Symes, C. B. 1940: Malaria in Nairobi. E. afr. med. J. 17, 291.

Teesdale, C. 1954: Freshwater moluscs in the coast province of Kenya with notes on an indigenous plant and its possible use in the control of bilharzia. E. afr. med. J. 31, 351.

Teesdale, C. 1961: The use of continuous low-dosage copper sulfate as a moluscicide on an irrigation scheme in Kenya. Bull. Wld Hlth Org. 25, 563.

Teesdale, C. 1962: Ecological observations on the moluscs of significance in the transmission of bilharziasis in Kenya. Bull. Wld Hlth Org. 27, 759 – 782.

Teesdale, C.; G. S. Nelson 1958: Recent work on schistosomes and snails in Kenya. E. afr. med. J. 35, 427 – 435.

Thompson, B. W.; H. W. Sansom 1967: Climate. In: Nairobi: City and Region. Ed. by W. T. W. Morgan. Nairobi, London, New York.

Vanek, E. 1976: Epidemiologisch-serologische Untersuchungen über das Q-Fieber in Kenia/Ostafrika. Habil. Schr. Ulm.

Vanek, E.; B. Thimm 1976: Q-fever in Kenya, Serological investigations in man and domestic animals. E. afr. med. J. 53, 678 – 684.

Verhagen, A. R. H. B. 1974: Leprosy. In: Vogel, L. C., A. S. Muller, R. S. Odingo et al. (Edit.), Health and Diseases in Kenya, East African Literature Bureau. 205 – 212. Nairobi.

Vogel, L. C. 1971: Implications of the transfer of basic health services from County Councils to Central Government in Kenya. In: Health and Disease in Africa. Proceedings of the East African Medical Research Council Scientific Conference 1970. Ed. G. Clifford Gould. East African Literature Bureau. 9, Nairobi.

Vogel, L. C.; A. G. Dissevelt; I. N. Van Luijk; M. Shimoni; A. Sjoerdsma; H. W'Oigo 1974: Operation research in outpatient services experiments in Kenya. Proceedings 9th Int. Congr. Trop. Med. & Mal. Athens, 14. 10. 1974, Athen.

Vogel, L. C.; P. O. Huma 1974: Yaws. In: Vogel, L. C., A. S. Muller, R. S. Odingo et al. (Edit.): Health and Diseases in Kenya, East African Literature Bureau, Nairobi, 371 – 374.

Vogel, L. C.; A. S. Muller; R. S. Odingo; Z. Onyango; A. de Geus (Edit.) 1974: Health and Diseases in Kenya. East African Literature Bureau, Nairobi.

Vogel, L. C.; A. C. Sjoerdsma; M. Shimoni; H. W'Oigo 1974: The use of drugs in an outpatient departement and ways and means to simplify pharmacy administration. In: The use and abuse of drugs and chemicals in tropical Africa. Proceedings of the E.A.M.R.C., Scientific Conference, Nairobi, 1973 Ed. Bagshawe, A. F., E. Afr. Lit. Bureau, 455 – 461.

Vogel, L. C.; H. O. W'Oigo; W. J. Swinkels; A. C. Sjoerdsma 1976: Cost analysis of outpatient services at Kiambu District Hospital, Kenya. E. afr. med. J. 53, 236 – 243.

Walton, G. A. 1950: Relapsing fever in the Meru District of Kenya. E. afr. med. J. 27, 94 – 98.

Walton, G. A. 1955: Relapsing fever in the Digo-district of Kenya-Colony (Kwale Distr.). E. afr. med. J. 32, 377 – 393.

Wamola, I. A.; J. N. Itotia; P. J. Say; Cruickshank 1974: Diarrhoeal diseases due to shigella and enteropathogenic Escherichia coli. In: Vogel, L. C., A. S. Muller, R. S. Odingo et al. (Edit.), Health and Diseases in Kenya, East African Literature Bureau. Nairobi, 185 – 192.

Webbe, G.; A. S. Msangi 1958: Observations on three species of bulinus on the east coast of Africa. Ann. trop. Med. Parasit. 52, 302.

Weigt, E. 1930/31: Die Kolonisation Kenias. Mitteilungen der Gesellschaft für Erdkunde zu Leipzig. 51, 25 – 123.

Weigt, E. 1955: Europäer in Ostafrika. Klimabedingungen und Wirtschaftsgrundlagen. Kölner Geographische Arbeiten 6/7. Köln.

Weigt, E. 1963: Wirtschaftliche und soziale Probleme der neuen Staaten Ostafrikas. Tijdschrift voor Economische en Sociale Geografie. 54, 229 – 237.

Wheeler, M. 1969: Medical manpower in Kenya and some of its implications. E. afr. med. J. 46, 93 – 101.

Who controls industry in Kenya? Report of a working party. Nairobi 1968.

Wijers, D. J. B. 1963: Studies on the vector of kala azar in Kenya. II. Epidemiological Evidence. Ann. trop. Med. Parasit. 57, 7 – 18.

Wijers, D. J. B. 1971: A ten years study of kala azar in Tharaka (Meru district, Kenya), I. Incidence studies from the records of Marimanti. E. afr. med. J. 48, 533 – 550.

Wijers, D. J. B. 1974: Filariasis. In: Vogel, L. C., A. S. Muller, R. S. Odingo et al. (Edit.), Health and Diseases in Kenya, East African Literature Bureau, Nairobi, 357 – 362.

Wijers, D. J. B. 1974: Leishmaniasis. In: Vogel, L. C., A. S. Muller, R. S. Odingo et al. (Edit.), Health and Diseases in Kenya, East African Literature Bureau, Nairobi, 331 – 338.

Wijers, D. J. B., D. M. Minter 1962: Studies on the vector of kala azar in Kenya. I. Entomological evidence. Ann. trop. Med. Parasit. 56, 462 – 472.

Wijers, D. J. B.; D. M. Minter 1966: Studies on the vector of kala azar in Kenya. V. The outbreak in Meru district. Ann. trop. Med. Parasit. 60, 11.

Wijers, D. J. B.; P. N. Munanga 1971: Schistosomiasis on Mfangano Island (South Nyanza, Kenya). E. afr. med. J. 48, 135 – 140.

Wijers, D. J. B.; S. Mwangi 1966: Studies on the vector of kala azar in Kenya. VI. Environmental epidemiology in Meru district. Ann. trop. Med. Parasit. 60, 373 – 391.

Wijers, D. J. B. 1977: Bancroftian filariasis in Kenya I. Prevalence survey among adult males in the Coast Province. Ann. trop. Med. Parasit. 1971, p. 313 – 332.

Wijers, D. J. B.; Kinyanjui, H. 1977: Bancroftian filariasis in Kenya II. Clinical and parasitological investigations in Mambrui, a small coastal town and Jaribuni, a rural area more inland (Coast Province). Ann. trop. Med. Parasit. 1971, p. 333 – 346.

Wijers, D. J. B.; Kiilu, G. 1977: Bancroftian filariasis in Kenya III. Entomological investigations in Mambrui a small coastal town and Jaribuni, a rural area more inland (Coast Province). Ann. trop. Med. Parasit. 1971, p. 347 – 360.

Willet, K. G. 1965: Some observations on the recent epidemiology of sleeping sickness in Nyanza Region, Kenya and its relation to the general epidemiology of gambian and rhodesian sleeping sickness in Africa. Trans. roy. Soc. trop. Med. Hyg. 59, 374 – 394.

Wykoff, D. E.; G. R. Barnley; M. M. Winn 1968: Leishmania donovani in Uganda. In: Uganda Atlas of Disease Distribution. Kampala.

Wykoff, D. E.; G. R. Barnley; M. M. Winn 1969: Studies on kala azar in Uganda entomological observations. E. afr. med. J. 46, 204 – 207.

Young, S.; A. Kondi; H. Foy: Anaemias 1974. In: Vogel, L. C., A. S. Muller, R. S. Odingo et al. (Edit.), Health and Diseases in Kenya, East African Literature Bureau, Nairobi, 431 – 436.

Ziedses Des Plantes, M.; A. R. H. B. Verhagen; D. Leiker; J. W. Koten 1968: Leprosy in Kenya. E. afr. med. J. 45, 371 – 377.

Illustrations

Acknowledgement

Photo 8: Frido J. W. Bader, Berlin
Photo 36: Dietrich O. Müller, Berlin
Photos 52 – 54, 58 – 60: H. J. Diesfeld, Heidelberg
all other photos: Hans K. Hecklau, Trier

Fig. 1. Kenya is a country of great regional diversity in the sense that the natural conditions of life of its population are extraordinarily varied. Kenya's palm-fringed coast along the Indian Ocean, with its fine-grade sand beaches and rich marine life, belong to the earth's most renowned holiday areas.
South of Mombasa 5. 1. 1976

Fig. 2. In the great national parks of the country the recollection of a pristine Africa, of the abundance of game in magically beautiful scenery, remains alive.
Tsavo National Park 29. 3. 1965

Fig. 3. The vast and gently undulating plains ... and the varied forms of the dry savanna, are the realm of pastoral nomads.
Close to Narok 5. 4. 1965

Fig. 4. Persevering and undemanding, the pastoral nomads remain to roam the areas of scant precipitation, always on the move to new pastures and waterholes, their existence constantly threatened by the loss of their herds when the rains fail for years on end, when grass and waterholes dry up.
Close to North Horr (East of Lake Rudolf) 12. 2. 1974

Fig. 5. What a contrast this presents to the fertile, well-watered and densely settled Kenya Highlands inhabited by the small farmers. Numerous cultivated plants from Africa and Asia, America and Europe flourish here—cereals and tubers, vegetables and fruit, coffee and tea. Healthy cattle thrive on the lush meadows.
Kisii Highlands 10. 4. 1965

Fig. 6. The rainforests of higher altitudes, with their many species, remain almost untouched, although economic demands have led to the conversion of more and more natural forests into agricultural land or plantations of pines and cypresses. Preparation for afforestation.
Aberdare Range 10. 2. 1966

Fig. 7. Pine forest (Pinus padua) along-side the road Nakuru-Eldoret. Altitude 9,109 feet 31. 12. 1976

Fig. 8. In the upper reaches the montane forests are fringed by bamboo . . . Mount Kenya 6. 2. 1966

Fig. 9. In the upper reaches the montane forests are fringed by bamboo and heather growth which gives way gradually to grassland and moorland. Mount Kenya 5. 2. 1966

118

Fig. 10. The Rift Valley system of East Africa is one of the most striking features of the earth's surface. In some places the shoulders of the rift rise more than a thousand meters above its floor.
Elgeyo Escarpment on the Eldoret-Tambach road. 5. 6. 1965

Fig. 11. In Kenya the formation of the Rift Valley has been accompanied by particularly marked volcanic activity, which has persisted in historic times. Lava from an outflow dating from the end of the last century. In the background youthful forms of extinct volcanoes.
South of the Chyulu Range. 3. 1. 1976

Fig. 12. However, Mount Kenya, the highest peak of which—Batian—reaches 5,199 m above sea level, is only a ruin of volcano, the more resistant vent-filling of which has been exposed from the surrounding volcanic masses by erosion.
Mount Kenya with Tow Tarn Col and Batian (17.058 F, right) and Nelion (17.022 F, left) 5. 2. 1966
Mount Kenya 6. 2. 1966

Fig. 13. Herdsmen belong to the conservative groups of the population which maintain the old traditions. Their extremely sparsely populated regions offer little chance to encounter innovations. The illustration here is of young Pokot girls, dressed in goatskins and richly adorned with decorative scars, bead necklaces, and brass rings. Both are illiterate. North of Cherangani Hills. 4. 1. 1967

Fig. 14. Pokot herdsman with spear, stool, traditional cloak, headdress and lip plug.
North of Cherangani Hills. 4. 1. 1976

Fig. 15. The farming population has accepted innovations to a much greater degree than the herding peoples. It is more effectively integrated into the money economy. Farming family in South Nyanza. Both the children attend school. Income from cash crops permits the family to acquire clothes and modest domestic items.
Homa Bay, 8. 4. 1965

120

Fig. 16. Sticking to old customs and traditions is also reflected in the settlements of the pastoral population. Detail of a Masai kraal in Narok District. 4. 4. 1965

Fig. 17. The cultural change among the farming population is demonstrated in the heterogeneous settlement pattern. Modern Kenya presents traditional circular huts with conical grass roofs, rectangular houses with grass roofs, and rectangular, solidly-built houses with corrugated iron roofs and divided into several rooms. Tick-borne diseases occur less frequently since ticks find conditions of life less congenial in modern houses than in the thatched huts. Settlement near Machakos, 16. 2. 1974

Fig. 18. A widely-used contemporary construction method; a framework of poles is covered with unburnt loam, the roof with corrugated iron. The houses are usually equipped with several rooms, windows and doors. Near Nyeri, 13. 2. 1966

Fig. 19. A characteristic Luo settlement, in which a few huts and storage bins are surrounded by a circular and very densely grown Euphorbia hedge. Cattle are kept within this fence during the night. In earlier times it used to serve as protection against raids.
On the Kavirondo Gulf, 30. 11. 1967

Fig. 20. Storage huts, a traditional feature of settlements almost everywhere in Kenya. They consist of wickerwork shaped like a large basked and are lined with mud. As a protective measure against dampness the store is set on a stage of branches; it is surmounted by a grass roof. A large part of the stored harvest rots or is infested with vermin.
Bungoma District, 21. 12. 1967

Fig. 21. Cassava is an important reserve foodstuff which can be left in the ground without perishing. An unbalanced diet made up only of cassava leads to the protein deficiency disease of kwashiorkor.
Bungoma District, 21. 12. 1967

Fig. 22. Progressive farmer's house on Mount Kenya at about 2,000 m above sea-level. It has several rooms, a corrugated iron roof, wooden doors, windows, and shutters. The mud walls are clad with bamboo. It is furnished in a European manner.
Meru District, 28. 1. 1966

Fig. 23. Stable and storehouse of timber, with a pit latrine in the front; feeding trough and yard for spraying cattle adjacent to it. Cattle are sprayed twice-weekly against disease-bearing ticks.
Meru District, 28. 1. 1966

Fig. 24. Among progressive farmers the cattle are fully integrated, i.e. fodder for the animals is produced on the farm and their dung is used on the fields. Under traditional African farming the cattle were left to graze in the bush beyond the farm holding.
Meru District, 28. 1. 1966

Fig. 25. Settlement scheme in Meru District. The farms are run according to the know-how of modern tropical agriculture. Monocultures are predominant, here chiefly maize, wheat, pyrethrum, potatoes, and vegetables among others.
Meru District, 29. 1. 1966

Fig. 26. Kikuyu farmsteads in the Aberdare Range, with maize as the main subsistence crop and coffee as the main cash crop
Nyeri District, 14. 2. 1966

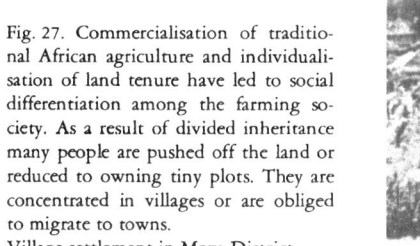

Fig. 27. Commercialisation of traditional African agriculture and individualisation of land tenure have led to social differentiation among the farming society. As a result of divided inheritance many people are pushed off the land or reduced to owning tiny plots. They are concentrated in villages or are obliged to migrate to towns.
Village settlement in Meru District, 28. 1. 1966

124

Fig. 28. In western Kenya sorghum and other millets are used to supplement maize as the staple food and provide the raw material for brewing beer.
Bungoma District, 28. 1. 1966

Fig. 29. Mixed cultivation occurs widely. In the foreground are seen sweet potatoes, followed by sorghum and cassava, with bananas in the background.
Bungoma District, 28. 1. 1966

Fig. 30. Tree cultivation characterises the coastal scene: coconut and cashew nut, citrus, mango, and papaya trees. Below the trees, maize, cassava, sorghum, etc., are grown.
Tiwi, south of Mombasa, 5. 1. 1976

Fig. 31. On the coast rectangular houses, roofed with palm leaves, predominate. The walls are made up of coral limestone, cemented together with loam.
Tiwi, 5. 1. 1976

Fig. 32. Maize is ground into posho between mill stones.
Tiwi, 5. 1. 1976

Fig. 33. Pestle used in the preparation of porridge.
Tiwi, 5. 1. 1976

Fig. 34. The most successful irrigation scheme at present is the Mwea Tebere Irrigation Scheme, which comprises 4.660 ha and is located in the southern extremities of Mt. Kenya at heights of about 1.200 m above sea-level (D 1 f).
Tenant havesting rice, 14. 2. 1966

Fig. 35. Whereas the large and well organised irrigation schemes are able to control the infestation of their populations by malaria and schistosomiasis, the smaller ones fail to do so. In weed-covered ditches both the agents of bilharziasis and the snails which act as intermediary hosts find ideal living conditions.
Taveta, 8. 3. 1974

Fig. 36. Termite hills in the dry savanna. Termites are caught, roasted and eaten. They are valued as a protein-rich food. The termite hills also house phlebotomes, the sting of which transmits leishmaniasis infections (kala azar).
North east of Mt. Elgon, 9. 4. 1967

Fig. 37. Overgrazing of natural vegetation by cattle, goats and sheep leads to the destruction of the vegetation and subsequent soil erosion. Goats and fat-tailed sheep in the vicinity of Taveta. Taveta District, 14. 2. 1966

Fig. 38. At most altitudes in Kenya soil erosion has already caused serious damage, especially in Machakos District. Machakos District, 14. 2. 1966

Fig. 39. Cattle are not kept according to a strictly economic standpoint, but serve among other things as status symbols. Herd of goats at a watering place in Narok District, north of Lake Magadi. Since men and animals make equal use of these watering places the water is not only muddy but also infested by numerous germs.
Narok District, 18. 3. 1974

128

Fig. 40. Numerous wells were dug by the administration in order to open up to the herding population those areas of the arid zone which did not possess natural watering places.
Kalacha, east of Lake Rudolf, 12. 2. 1974

Fig. 41. Women's tasks are not confined to the cultivation and preparation of food for the family, but also include provision of water, which is often procured a long distance away.
Taveta, 9. 3. 1974

Fig. 42. Babies are carried on their mothers' backs until the birth of the next child. The loss of the mother's milk and displacement from the warm and sheltered place on the mother's back in favour of new arrivals frequently lead to nutritional disorders and colds, often with fatal consequences.
Taveta, 9. 3. 1974

Fig. 43. It is difficult to involve small-holders in production for the market. The quantities offered for sale by the farmers are often so small that they can be taken to the market or collecting point on the head. The amounts harvested vary. In good years the farmers sell maize surplus to the requirements of their own family.
Chamasiri Market, Bungoma District, 21. 12. 1967

Fig. 44. Farmers who cultivate coffee are required to be members of a Coffee Growers' Society, which offers advice, supplies plants, processes, and markets the coffee berries.
Coffee factory of a Coffee Growers' Society in Meru District. 27. 1. 1966

Fig. 45. Valuable export crops like coffee, tea and pyrethrum do not grow in farming districts at altitudes below 1,500 m. As for sales, the farmers are restricted to marketing the surplus production of subsistence crops and have to confine themselves to the cultivation of low value cash crops like sisal and cotton; the latter is carried to the collection points in small quantities. From there the Lint Marketing Board takes over. Extended transport distances increase the costs of production.
Bungoma District, 21. 12. 1967

Fig. 46. The Large Farm Areas in the highlands play an important role in supplying the non-agricultural population of Kenya. Wheat, maize, malting barley and other crops are grown within the system of ley farming.
Wheat fields near Molo, 30. 1. 1967

Fig. 47. Only about a quarter to a third of the arable land of farms in the Large Farm Area is utilised for tillage. The remainder serves as grazing land for dairy and beef cattle.
Sheep breeding is of secondary importance. Herd of cattle near Molo, 30. 1. 1967

Fig. 48. In the non-arable areas large-scale farms practice pasture-farming, chiefly the raising of beef cattle. Dairy cattle are only kept in the proximity of markets and in ecologically favourable areas.
Herd of beef cattle near Nanyuki, 1. 2. 1966

Fig. 49. Grape-fruit trees on irrigated land in Taita District. 12. 3. 1974

Fig. 50. The largest compact area of tea plantations in Kenya, totalling some 35,000 ha., is situated in the Kericho region.
Tea pickers at work.
Kericho, 31. 12. 1975

Fig. 51. The families of agricultural labourers employed on pastoral farms, mixed farms, and plantations live in small closed settlements. Only in larger establishments are these settlements equipped with the necessary social facilities.
Settlement for agricultural labourers in the vicinity of Kericho, 31. 12. 1975

Fig. 52. Government hospital in Marsabit.
Out-patients
Marsabit 13. 2. 1974

Fig. 53. Government hospital in Marsabit
The wards
Marsabit 13. 2. 1974

Fig. 54. Garissa hospital
The wards
Kinango 19. 2. 1974

Fig. 55. Kinango Mission Hospital. Reception.
Kinango, 7. 3. 1974

Fig. 56. Kenya has at its disposal a partially well-developed system of central places, which meet the requirements of the agrarian "Umland". Main shopping street in Nakuru, the first-order centre for the Large Farm Areas.
Nakuru, 31. 12. 1075

Fig. 57. Besides the central places there are many small trading centres where dukas (small retail outlets) offer goods for daily consumption to the rural population. All sorts of medicine, too, can be obtained in these dukas.
Duka near Kitui, 16. 2. 1966

134

Fig. 58. Kenyatta National Hospital
Outpatient Department
Nairobi 15. 3. 1974

Fig. 59. Kenyatta National Hospital
Wards
Nairobi 15. 3. 1974

Fig. 60. Medical Research La
Ministry of Health
Nairobi 15. 3. 1974

Additional material from *Kenya*

ISBN 978-3-642-66937-8 (978-3-642-66937-8_OSFO2),

is available at http://extras.springer.com